Memoirs of the Museum of Anthropology
University of Michigan
Number 47

Elamite and Achaemenid Settlement on the Deh Lurān Plain

Towns and Villages of the Early Empires in Southwestern Iran

edited by
Henry T. Wright
and James A. Neely

with a preface by Frank Hole
and contributions by
Elizabeth Carter, Piotr Michalowski,
Pierre de Miroschedji, James A. Neely, and Henry Wright

Ann Arbor, Michigan
2010

©2010 by the Regents of the University of Michigan
The Museum of Anthropology
All rights reserved

Printed in the United States of America
ISBN 978-0-915703-72-2

Cover design by Katherine Clahassey

The University of Michigan Museum of Anthropology currently publishes two monograph series, Anthropological Papers and Memoirs, as well as an electronic series in CD-ROM form. For a complete catalog, write to Museum of Anthropology Publications, 4013 Museums Building, 1109 Geddes Avenue, Ann Arbor, MI 48109-1079, or see www.lsa.umich.edu/umma/publications

Library of Congress Cataloging-in-Publication Data

Elamite and Achaemenid settlement on the Deh Luran Plain : towns and villages of the early empires in southwestern Iran / edited by Henry T. Wright and James A. Neely ; with a preface by Frank Hole ; and contributions by Elizabeth Carter ... [et al.].
    p. cm. -- (Memoirs of the Museum of Anthropology, University of Michigan ; no. 47)
  Includes bibliographical references.
  ISBN 978-0-915703-72-2 (alk. paper)
 1. Deh Luran Plain (Iran)--Antiquities. 2. Elam--Antiquities. 3. Achaemenid dynasty, 559-330 B.C. 4. Deh Luran Plain (Iran)--History, Local. 5. Cities and towns, Ancient--Iran--Deh Luran Plain. 6. Villages--Iran--Deh Luran Plain--History. 7. Pottery, Ancient--Iran--Deh Luran Plain. 8. Excavations (Archaeology)--Iran--Deh Luran Plain. 9. Land settlement patterns--Iran--Deh Luran Plain. 10. Canals--Iran--Deh Luran Plain--History. I. Wright, H. T. (Henry T.) II. Neely, James A. III. Carter, Elizabeth, 1943-
  DS324.D44E43 2010
  935--dc22
                       2010018925

The paper used in this publication meets the requirements of the ANSI Standard Z39.48-1984 (Permanence of Paper)

Contents

LIST OF ILLUSTRATIONS		*v*
LIST OF TABLES		*vii*
PREFACE, *by Frank Hole*		*ix*

1 INTRODUCTION   *1*
    by Henry T. Wright and James A. Neely
        The Dynamics of States and Empires in Early Historic Mesopotamia   *1*
        Geography of the Deh Lurān Plain   *2*
        A Brief History of Archaeological Research on the Deh Lurān Plain   *3*
        Plan of the Present Work   *6*
        Acknowledgments   *6*

2 RESEARCH DESIGN AND METHODS   *7*
    by James A. Neely

3 CERAMIC PHASE INDICATORS IN SURFACE ASSEMBLAGES   *11*
    by Elizabeth Carter and Henry T. Wright
        The Later Early Dynastic Phase   *11*
        The Šimaški Phase   *14*
        The Sukkalmaḫḫu and Middle Elamite Phases   *14*
        The Neo-Elamite and Achaemenid Phases   *15*

4 THE ARCHAEOLOGICAL SITES AND THEIR INTERPRETATION   *23*
    by James A. Neely and Henry T. Wright

5 EARLY HISTORIC SETTLEMENT PATTERNS ON THE DEH LURĀN PLAIN   *83*
    by Henry T. Wright
        The Later Early Dynastic Phase   *83*
        The Šimaški Phase   *84*
        The Sukkalmaḫḫu Phase   *87*
        The Earlier Middle Elamite Phase   *89*
        The Later Middle Elamite Phase   *91*
        The Neo-Elamite Phase   *91*
        The Achaemenid Phase   *91*

6 WATER MANAGEMENT DURING THE LATER EARLY DYNASTIC, ELAMITE, AND ACHAEMENID PERIODS ON THE DEH LURĀN PLAIN   *95*
    by James A. Neely

7	TEXTUAL DOCUMENTATION OF THE DEH LURĀN PLAIN: 2550–325 B.C.	*105*
	The Mid–Late Third Millennium B.C. on the Deh Lurān Plain	*106*
	by Piotr Michalowski and Henry T. Wright	
	The Early to Mid-Second Millennium B.C.	*109*
	by Henry T. Wright	
	The Early First Millennium B.C.	*110*
	by Pierre de Miroschedji and Henry T. Wright	
	The Mid-First Millennium B.C.	*111*
	by Henry T. Wright	
8	A LAND BETWEEN: REFLECTIONS ON EARLY HISTORIC DEH LURĀN	*113*
	by Henry T. Wright and James A. Neely	
	Deh Lurān in the Trans-Regional Context	*113*
	Transforming Water Management and Food Production on the Deh Lurān Plain	*114*
	Reorienting the Deh Lurān Settlements around Corridors of Communication	*115*
	The Implications of this Survey for Future Archaeological Work	*116*

APPENDIX A. COUNTS OF ARTIFACTS IN SURVEY COLLECTIONS *117*
APPENDIX B. ARCHAEOLOGICAL SITES ON THE DEH LURĀN PLAIN *123*

REFERENCES CITED *125*

COLOR PLATES *133*

Illustrations

Figures

1.1. Greater Mesopotamia and the Deh Lurān Plain, *4*
1.2. Natural environments of the Deh Lurān Plain, *5*

2.1. Survey coverage of the Deh Lurān Plain, *8*

3.1. Ceramic vessels of the Early Dynastic III phase, *13*
3.2. Ceramic vessels of the Šimaški phase, *17*
3.3. Ceramic vessels of the Sukkalmaḫḫu and Middle Elamite phases, *19*
3.4. Ceramic vessels of the Neo-Elamite and Achaemenid phases, *21*

4.1. Map of Tāfuleh, *24*
4.2. Map of Musiyān, *27*
4.3. Cups and bowls of the IIIrd millennium B.C. from Musiyān, *29*
4.4. Bowls of the IIIrd millennium B.C. from Musiyān, *31*
4.5. Jar sherds of the IIIrd millennium B.C. from Musiyān, *33*
4.6. Jar sherds of the IIIrd millennium B.C. from Musiyān, *35*
4.7. Varia from Musiyān, *37*
4.8. Late IIIrd and IInd millennium B.C. goblets, bases, and bowls from Musiyān, *39*
4.9. Late IIIrd and IInd millennium B.C. jars from Musiyān, *41*
4.10. Late IIIrd millennium B.C. jars and basins from Musiyān, *43*
4.11. IInd millennium B.C. basin and bowl sherds from Musiyān, *45*
4.12. The distribution of IIIrd to Ist millennium B.C. ceramics on Musiyān, *46*
4.13. Map of Baulah, *47*
4.14. Late IIIrd and IInd millennium B.C. jars, basins, and bases from Tepe Baulah, *49*
4.15. Map of Tenel Ramon, *50*
4.16. IInd millennium B.C. jars and bases from Tenel Ramon, *53*
4.17. Jar sherds and figurine from Tenel Ramon, *55*
4.18. Basins or large jars of the IInd millennium B.C. from Tenel Ramon, *56*
4.19. Map of Tepe Farukhābād showing Sukkalmaḫḫu revetment, *57*
4.20. Map of Tepe Gārān, *59*
4.21. Cup, bowl and jar sherds of the IIIrd millennium B.C. from Tepe Gārān, *61*
4.22. Goblets, jars, bowls, and basins of the IInd millennium B.C. from Tepe Gārān, *63*
4.23. Basin and bowls of the mid-Ist millennium B.C. from Tepe Gārān, *65*
4.24. Jars of the mid-Ist millennium B.C. from Tepe Gārān, *67*
4.25. Map of Tepe Patak, *69*
4.26. Goblet, bowl, and jar sherds from Tepe Patak, *71*
4.27. Basins and large jars from Tepe Patak, *73*
4.28. Bowls and jars of the mid-Ist millennium B.C. from Tepe Patak, *75*
4.29. Map of DL-41, *77*
4.30. Map of Tepe Sohz, *78*
4.31. Map of Tepe Mohr, *79*
4.32. Late IIIrd millennium bowl, jar, and basins from Tepe Mohr, *80*

5.1. Settlements of the Later Early Dynastic phase, *85*
5.2. Settlements of the Šimaški phase, *86*
5.3. Settlements of the Sukkalmaḫḫu phase, *88*
5.4. Settlements of the Earlier Middle Elamite phase, *90*
5.5. Settlements of the Neo-Elamite phase, *92*
5.6. Settlements of the Achaemenid phase, *93*

6.1. Later Early Dynastic phase canals, *98*
6.2. Šimaški phase canals, *99*
6.3. Sukkalmaḫḫu phase canals, *100*
6.4. Earlier Middle Elamite phase canals and possible *qanāts*, *101*
6.5. Neo-Elamite phase canals and possible *qanāts*, *102*
6.6. Achaemenid phase canals, *qanāts*, and roads, *103*

B1. Map of archaeological sites on the Deh Lurān Plain, *124, inside back cover*

Color Plates

Pl. 1. Deh Lurān landscapes, *133*
Pl. 2. Images of Musiyān (DL-20), *134*
Pl. 3. Images of Farukhābād (DL-32), *135*
Pl. 4. Images of Gārān (DL-34), *136*
Pl. 5. Images of Patak (DL-35), *137*
Pl. 6. Images of DL-41, and new canal near Gārān, *138*
Pl. 7. Deh Lurān water management features, *139*
Pl. 8. Human and divine representations from the Deh Lurān Plain, *140*

Tables

5.1. Later Early Dynastic settlements, *84*
5.2. Šimaški settlements, *84*
5.3. Sukkalmaḫḫu settlements, *87*
5.4. Earlier Middle Elamite settlements, *89*
5.5. Neo-Elamite settlements, *92*
5.6. Achaemenid settlements, *92*

6.1. Proposed formal *qanāts* and canals on the Deh Lurān Plain, *97*

A1. Musiyān Jemdet Nasr EDI–EDII ceramics, *118*
A2. Musiyān Elamite ceramics, *119*
A3. Tenel Ramon Elamite ceramics, *120*
A4. Tepe Gārān Elamite and Achaemenid ceramics, *121*
A5. Tepe Patak Elamite and Achaemenid ceramics, *122*

Preface

Frank Hole

Archaeological exploration of the Deh Lurān Plain began in the winter of 1902–1903 when a French team, under the direction of Gautier and Lampre, excavated for two months at Tepe Musiyān, and tested several other visible mounds (Plate 1). Choosing Musiyān, the largest site on the plain, the French constructed a camp fortified by a ditch and rampart, and hired some 140 local tribesmen. After this short season, described in a lengthy report (Gautier and Lampre 1905), archaeological activity ceased on the Deh Lurān Plain for nearly sixty years. Just before the First World War, an exploration company detonated seismic soundings in the search for oil, and constructed a narrow gauge railway to carry tar to Amarah (in what is today Iraq) from the large seep near the base of the mountains. In 1959, Robert Braidwood and Herb Wright passed through Deh Lurān on a brief reconnaissance and visited Tepe Mohammad Jaffar, a site at which the French had found early pottery and lithics. In 1961 and 1963, Kent Flannery and I excavated at this site, now known as Ali Kosh.

In September 1961, driving an old Army jeep that had already seen severe service, we boarded a wooden ferry, attached by a cable to either side of the Kharkeh River. We were joined by a herd of sheep, some donkeys, and herd boys, while the boatman cranked the truck engine to sputtering life and engaged the drum that slowly wound the cable and dragged us across the river. On the far side, a narrow gravel road led us across a seemingly endless steppe whose principal signs of habitation were abandoned camps of transhumant herders. After crossing several dry wadis, we finally arrived at the Dawairij River. The ankle-deep water then gave no hint of the winter floods that in later years would isolate our camp for weeks on end.

There were few places in Iran more remote from the advances of technology than Deh Lurān. The three permanent settlements survived with strictly subsistence economies, supplemented by smuggling across the border with Iraq. Sometime before Kent and I arrived in the fall of 1961, the Iranian Army had established a small post and an electric generator to provide lighting in this sensitive border area. Each house in the village of Deh Lurān was equipped with an electric line and bulb, but the stress on the generator proved too great, the generator failed, and electricity was only a dim memory. Our representative arranged for us to be housed in the local two-room school in the village. As we discovered to our dismay, it was only a short distance from a spring whose sulphurous vapors descended nightly into town with suffocating effect.

The Deh Lurān Plain was largely untouched by agriculture or even vehicle tracks, and herds of gazelles grazed the steppe vegetation. While we were able to secure some rounds of barley bread from a local, no other food was available so we were persuaded to use our jeep to hunt a gazelle. One of the village men had a shotgun and, with him riding in the front seat, gun cocked, we set off in pursuit of a gazelle.

Knowing that they run in a semicircle when chased, I attempted to cut them off so we could get in a good shot. The plain was flat and inviting and we sped across it, bearing down on the herd. As we closed on a gazelle, the man pulled the trigger, but the shell misfired and before he could reload, the gazelle suddenly disappeared. Too late I realized that they had plunged into a wadi and before I could brake the jeep, we too had plunged in, nose down. The gazelle were long gone, as was our radiator, which leaked profusely thereafter.

Our first survey was carried out dashing from visible site to visible site, carrying a bucket that we set under the radiator while we got out to inspect the sites. The radiator was only part of the problem. There was no fresh water in Deh Lurān. Beneath the school was a cistern that collected rainwater in the winter, but in September it was nearly depleted; it sufficed only for cups of tea. Until the rains would fall in November, the natives made do with the brackish water from the spring, or water that was irregularly trucked in from a freshwater spring some 15 miles away and dumped into a tank in the middle of the village.

Because of the absence of consumable food and the sorry condition of our radiator, rather than backtrack we decided to move to Mehran, the next town to the northwest along the mountain front. I patched the radiator with some tar from the roof of the school and we set off to an uncertain fate. Well before we reached Mehran, the radiator's patch gave out and we had to stop regularly to add water. In Mehran, we asked for food and drink but were able to secure only a four-ounce bottle of lemon extract, which I, feverish and dry, sucked on while Kent drove up the mountain to Ilam. This inauspicious start on survey did nothing to detract from our enthusiasm to return in cooler weather, with a good vehicle and plenty of supplies.

Despite the physical problems, we felt lucky to have found some interesting sites, including Ali Kosh, which we returned to test in December and excavated in 1963, along with Tepe Sabz. While these excavations began to give us a sequence of early settlements, we realized that we needed to gain a complete picture of settlement of all periods. This led us to carry out an intensive survey in 1969, headed by Jim Neely, in tandem with our excavations at Chagha Sefid, which provided a stratigraphic link between the latest layers of Ali Kosh and the earliest layers of Tepe Sabz. Using aerial photos kindly supplied by Herb Wright, and a copy of an old British Army topographic map, Jim, helped primarily by Pierre de Miroschedji and Nathalie Desse-Berset (both lent by Jean Perrot's French expedition to Susa), and Elizabeth Carter, carried out an intensive survey of as many visible sites as possible. A volume, *Early Settlement and Irrigation on the Deh Lurān Plain* (Neely and Wright 1994), describing the prehistoric periods has already appeared, and this volume carries the study up to the end of Achaemenid times.

Comparison of the aerials we used with "Google Earth" today indicates just how drastically the landscape has changed. While Kent and I could drive without impediments—other than an occasional gully—and our tracks were visible years later, today the entire plain is under irrigation. Ancient sites, canals, check dams, and terraces have been destroyed by leveling and robbing of stones for other constructions. A military base on the site of Chagha Sefid has rendered it unsuited to further excavation. The little settlement of Deh Lurān is now many times its previous size. Only careful ground survey could reveal how much archaeology was lost when Deh Lurān was a battleground during the Iraq-Iran war.

Despite its obscure location, now less so because of the building of roads and bridges, and access to markets and education, the Deh Lurān Plain plays a central and critical role in the archaeology of western Iran. It, along with the Susiana Plain, comprises the best-documented regions of the Iranian lowlands. The sites of Ali Kosh, Chagha Sefid, Tepe Sabz, and Farukhābād played an important role in revealing the history of agriculture and animal husbandry and the relations between highland and lowlands. Deh Lurān is on the ancient Achaemenid road that connected Susa to Sardis in western Anatolia. Deh Lurān was also seasonal pasture for Arab tribes from Mesopotamia, and Luri and Kurdi tribes from Iran. Today, Deh Lurān sits on an oil reservoir, as yet unexploited, and it has become one of the prime agricultural regions as a result of the building of dams and reservoirs, which supply irrigation. Considering the poor quality of river water and the underlying gypsum layers, it remains to be seen whether this intense irrigation can be sustained.

Deh Lurān has had a rich history, one that stretches back some tens of thousands of years, to judge from finds of Paleolithic artifacts, yet it is known primarily for its prehistory. One day, this picture could be importantly augmented by further excavation of later sites, which will reveal more insight into the ebb and flow of Deh Lurān's prosperity that is revealed in the survey results reported here. This volume is a snapshot of history, a glimpse of what might be revealed through intensive excavation.

Chapter 1

Introduction

Henry T. Wright and James A. Neely

The settlements of the Deh Lurān Plain played a small role in the succession of early states and empires that dominated Southwest Asia from 2550 to 325 B.C. (Fig. 1.1). Nonetheless, the plain was located in a key position in the nexus of routes that connected competing polities and integrated transregional empires, the most important being the foothill road connecting Susa with the Diyala and Upper Mesopotamia. This later became part of the Achaemenid Royal Road from Susa to Sardis. Both integrative and competitive processes are documented by changing settlement size and location. No less important and interesting, however, were general changes in later Holocene climatic patterns toward more arid conditions and related changes in river regime on the plain, and the changes in agriculture, herding, and resource extraction that Deh Lurān's occupants devised to sustain themselves in such circumstances.

Earlier Deh Lurān excavations and studies evaluated hypotheses about the change from food collecting to food production and the initial development of irrigation techniques in early Holocene times before 4500 B.C. (Hole, Flannery, and Neely 1969; Hole 1977). Others focused on the relation between trade and political developments from 4500 to 2400 B.C. (Wright ed. 1981). Our previous monographic report on our settlement survey (Neely and Wright 1994) focused on changing population, settlement hierarchy, and agriculture as village societies developed into states between 6000 and 2400 B.C. This monograph is concerned with a marginal province in successive transregional polities, Deh Lurān's settlement patterns providing indices of the degree of investment in agriculture and transport from 2550 B.C. to 325 B.C.

Unfortunately, the small excavations on the plain do not provide a complete stratigraphic sequence of the period under consideration here. The excavations at Farukhābād (Wright 1981) provide a relatively continuous stratigraphic record only up to the middle of the IInd millennium B.C. Excavations at Tepe Gārān, which would document the local ceramics of the late IInd and Ist millennium B.C., have not yet been undertaken. However, we can date ceramics from this long period of time by reference to excavated samples from Susa to the southeast and from alluvial Mesopotamia proper to the south and west. With the ceramic evidence, we can date the occupations of sites located by our surface survey, as well as traces of canals and other cultural features, and define the settlement pattern of each period. With these data and with the available evidence of contemporary documents in Sumerian, Akkadian, Greek, and other languages, we can evaluate ideas about economic and political trajectories during this long period of time. This monograph presents a detailed summary of the evidence from the plain relevant to periods between 2550 and 325 B.C., together with our reassessment of changing regional patterns, so that others may use these data to answer new questions. This is the second of three planned reports. The third will continue the study up to the end of the Early Islamic period with the impact of the Mongol Invasion around A.D. 1250.

The Dynamics of States and Empires in Early Historic Mesopotamia

The view of empires developed by scholars in the historical sciences is grounded in the later empires with copious written documentation and striking material manifestations. Our vision owes much to our knowledge of traditional China and Persia, ancient Egypt and Rome, and Mexico and Peru at the time of the

Spanish conquest. We see empires as large transregional polities whose elites dominated their neighbors; appropriated most of the resources they desired; built systems of communication, bureaucratic control, and military domination; and promulgated ideologies of universal philosophy, law, and sovereignty. We can suggest that early empire builders would not know how to bring all these elements together, but until recently we had little knowledge of early efforts to build stable empires in worlds of competing smaller states. Recent research on the cuneiform documents of ancient Mesopotamia has begun to provide us with evidence of the rise, operation, and disintegration of some of the first empires in the region and, indeed, in the world.

This monograph deals with a district at times a frontier for and at other times an interstitial area within a succession of emergent larger polities. It is thus relevant to issues of frontier maintenance and internal communication. This is, however, the first study of this phase of culture history in this area. Future excavation and study of in situ cuneiform archives will allow researchers to make useful contributions to the broad issues of imperial culture process. We will limit ourselves to presenting the evidence from our survey with a few general remarks in Chapters 7 and 8.

Geography of the Deh Lurān Plain

The ridges and valleys of the Zagros front ranges formed during the Tertiary epoch as the Arabian fragment of the African continental plate impacted the Eurasian landmass, compressing and folding the sediments of the ancient Tethys Sea. The depression in front of the folded and uplifted outer Zagros has filled with sediments eroded from the Zagros and carried down by the tributaries of the Euphrates, Tigris, Karkheh, and Karun Rivers. As the glaciers melted, the sea rose to fill the lower, eastern end of this depression, reaching modern sea level before 3000 B.C. and creating the Persian Gulf.

The geography of the Deh Lurān Plain is treated in detail in the earlier publications noted above. However, a brief summary will be useful to those readers who do not have these on hand. The northeast edge of the Tigris-Euphrates alluvium is marked by the Jebel Hamrin, which reaches a maximum elevation of only 400 m. The Deh Lurān Plain is in a small trough between the Jebel Hamrin and the first great fold of the Zagros in the region of Pusht-i Kuh, the oak-covered Kuh-i Siah, a fold of limestone and sandstone rising to an elevation of nearly 1400 m. Northeast of this fold is an almost continuous scarp of soft Tertiary sediments, the upper surface of which is a plateau of gypsum at about 1300 m. The plain itself (Fig. 1.2), including its alluvial fans, averages almost 20 km in width from the Jebel Hamrin on the southwest to the foot of the Kuh-i Siah on the northeast. It extends for almost 60 km from northwest to southeast below the mountain front. The valley floor covers approximately 940 km^2. The topography of the plain has been formed by the action of two small rivers. The Mehmeh cuts through the scarp and enters the northwest corner of the plain, flows southeast to the middle of the plain, then turns abruptly southwest, cuts across the Jebel Hamrin, and disappears into the Tigris alluvium. The Dawairij or Āb-Dānān River cuts through the scarp and enters the northeast edge of the plain near its midpoint, then turns southeastward and flows out the southeast end of the plain. These rivers, which were aggrading, depositing sediment and building up the plain during much of early and middle Holocene times, cut their beds down at some point during the IInd millennium B.C., creating the entrenched channels in which they flow today (Kirkby 1977). This change in river regime, and human responses to it, will be a theme of the present work.

Characteristic vegetation associations allow one to divide the plain into four distinct areas, each of which presents different opportunities to farmers and herders (Hole, Flannery, and Neely 1969:16–19).

(1) The alluvial slopes had dense grassy vegetation on gravel-covered slopes where the rivers enter the plain, and a cover of sparse grass and *Zizyphus* or jujube trees elsewhere. These areas ordinarily do not retain enough moisture for rainfall agriculture, but various forms of hill slope terracing, the use of check dams in wadis, and small-scale irrigation with the water of various springs have allowed some farming in the past. Traditionally, transhumant pastoralists have used these slopes for grazing sheep and goats.

(2) The older alluvial plain formed by the fans of the two rivers, with denuded silty and sandy soils, has patches of such shrubs as *Prosopis*, a small legume, and *Alhaghi*, or camel thorn, a salt-tolerant succulent. These areas can retain moisture sufficient for dry farming in one year out of two, especially in areas where the water table is higher or where floodwaters can be concentrated. Floods late in the growing season, however, can easily destroy crops in such areas.

(3) The present river floodplains, now incised four to six meters below the surface of the alluvial fans, are subject to frequent flooding, and consequently have irregular surfaces with gravel and sand banks as well as channel scars. Dense thickets of *Tamarix*, the salt-cedar, and *Glycyrhrhyza*, the wild licorice, thrive because of the high water table.

(4) The saline depressions within the older alluvial plain are flooded in the winter and become dry salt-flats in the summer. A range of salt-tolerant shrubs live in this environment.

The wild fauna from contexts of the early IInd millennium B.C. at Farukhābād (Redding 1981) indicate something of the environment early in the span of time considered in this study. Soft-shelled turtle is common. Its presence implies a body of fresh water nearby, probably the Mehmeh River. Gazelle would have lived in uncultivated areas of steppe vegetation and if the equid remains are wild onager and not horses or donkeys (and the former are definitely attested), these too would have lived on open steppe. If the pig remains were from wild individuals, they would have lived in thickets along the rivers or in the saline depressions. Other wild fauna—hyenas, foxes, falcons, and doves—live in a variety of habitats.

Travel from alluvial Mesopotamia to Susiana and into the Iranian plateau are attested in the itineraries of military expeditions and government messengers beginning during the time of Akkadian control, about 2300 B.C. These records must be considered in relation to the location of known sites and to the traditional routes of access facilitated by topographic considerations (see Chapter 7). The major route of access is that traversing the axis of the plain along the mountain front, between Susa to the southeast to Der to the northwest. Several minor routes branch from this southwest toward alluvial Mesopotamia or northeast into the Zagros.

While we lack paleoecological studies of the early historic period on the Deh Lurān Plain, we are fortunate that from the Saimarreh Valley just 60 km north of Deh Lurān, we have new paleoecological information relevant to the IIIrd to Ist millennium people. Sixty kilometers northeast of Deh Lurān in a karstic depression on the contorted surface of the Saidmarreh landslip (Watson and Wright 1969) is Lake Mirabad (lat. 33°05'00" N/47°42'09" E, approximately 750 m ASL). Pollen from a 7.2-m core raised by Herbert Wright in 1963 has been studied by Van Zeist and Bottema (1977). The core was recently re-dated with improved ^{14}C techniques and the pollen evidence was supplemented with studies of other indicators by teams under the direction of Stevens et al. (2006) and Griffiths et al. (2001). In brief, these studies show that from about 4450 B.C. to about 2550 B.C., the proportions of pistachio and saltbush pollen decrease while oak pollen increases markedly, indicating a more closed and humid oak forest around the lake. Also, the proportion of ^{18}O indicates both winter and spring rainfall and the ratio of strontium to calcium in one major ostracod taxa decreases about 4200 B.C., indicating more humid conditions in general. There is a brief increase in pistachio after 2450 B.C., followed by a near disappearance of pistachio and a proportional drop in oak pollen, and a proportional increase in the pollen of grasses as well as *Plantago lanceolata*, normally considered a weed of overcultivation and overgrazing, reaching a peak about 1800 B.C. At the same time, the proportion of ^{18}O indicates a decrease in spring rainfall, and the ratio of strontium to calcium indicates more arid conditions. Note that it is possible that oak and pistachio remained constant while grasses and *Plantago* increased, but that with the proportional study of pollen percentages, an illusion of an oak decrease has been created. A study of pollen density and an estimation of pollen influx into the sediments, which could control for this problem, would require the extraction of a new core. In any event, the change in *Plantago* may indicate only local disturbance in the immediate Mirabad area. What is important is that after 2450 B.C., conditions fluctuate. However, from 1800 B.C. until the beginning of the common era, both the pollen and the proportion of ^{18}O indicate that conditions very much like those of today predominated.

Kirkby (1977) has presented evidence that the rivers on the Deh Lurān Plain cut down into the alluvial plain during the IInd millennium B.C. It seems likely that the initiation of downcutting was related to the shift to modern climatic conditions between 2550 and 1800 B.C.

A Brief History of Archaeological Research on the Deh Lurān Plain

J.-E. Gautier and G. Lampre of the Mission Archéologique en Perse first surveyed the plain in 1903. They followed the path of an earlier survey party led several years before by the director of the mission, Jacques de Morgan, westward through the Pusht-i Kuh region toward Kermanshah. Their main task was to investigate the large mound of Tepe Musiyān, where their excavations revealed architecture and ceramics of the Sukkalmaḫḫu Elamite phase, and of earlier phases. They also excavated the nearby cemetery of Tepe Aliabad and visited a number of other sites on the central portion of the plain, providing a brief description and a map (Gautier and Lampre 1905:60–62, Fig. 94). The observations of Gautier and Lampre are cited under the relevant site descriptions in Chapter 4.

Nearly sixty years later, the Director of the University of Chicago's Prehistoric Project, Robert J. Braidwood, and one of the project's geologists, Herbert Wright, visited the Deh Lurān Plain to examine possible early village sites involved in the process of plant and animal domestication. In the following year, Frank Hole (as recalled in his Preface to this volume) and Kent Flannery visited the plain to survey and to conduct test excavations at the site that is today called "Tepe Ali Kosh." Since Hole and Flannery did not have copies of Braidwood's survey notes, they assumed that he had recorded fewer than twenty sites, and gave the fourteen sites they visited on the central plain numbers ranging from DL-20 to DL-33 (Hole and Flannery 1962). In 1963, Hole instituted the Rice University Archaeological Project and returned with Flannery and James A. Neely to conduct major excavations at Tepe Ali Kosh (DL-21) and the later site of Tepe Sabz (DL-31), relevant to the beginnings of irrigation agriculture (Hole, Flannery, and Neely 1969). No formal survey was done. Early in 1968, Henry T. Wright of the University of Michigan Museum of Anthropology came to the plain for a season of excavation at Tepe Farukhābād (Wright 1981). The Farukhābād team visited a number of sites, but permanent site numbers were not given. Instead, most of these sites were revisited in 1969 by Neely's team, and these were entered into Neely's 1969 number series. All of the information acquired in 1960, 1961, 1963, and early 1968 has been incorporated without specific citation into the descriptions in Chapter 4. The survey project undertaken by Neely, then established at the University of Texas in Austin, in late 1968 and early 1969 was the first actually directed at the full recording of regional settlement patterns, as he details in the next chapter. Both Neely's survey and the simultaneous excavations of Hole at Chagha Sefid (DL-23) (Hole 1977) were supported by a grant from the National Science Foundation (GS-2194). Hole's strong support throughout our project is gratefully acknowledged. The survey began on November 1, 1968, and concluded on March 30, 1969. At various times, Neely was joined in the field by Anne and Michael Kirkby, the project's geographical specialists; Lynn Berry Fredlund from the University of Colorado; Nathalie Desse-Berset and Pierre de Miroschedji visiting from the Délégation

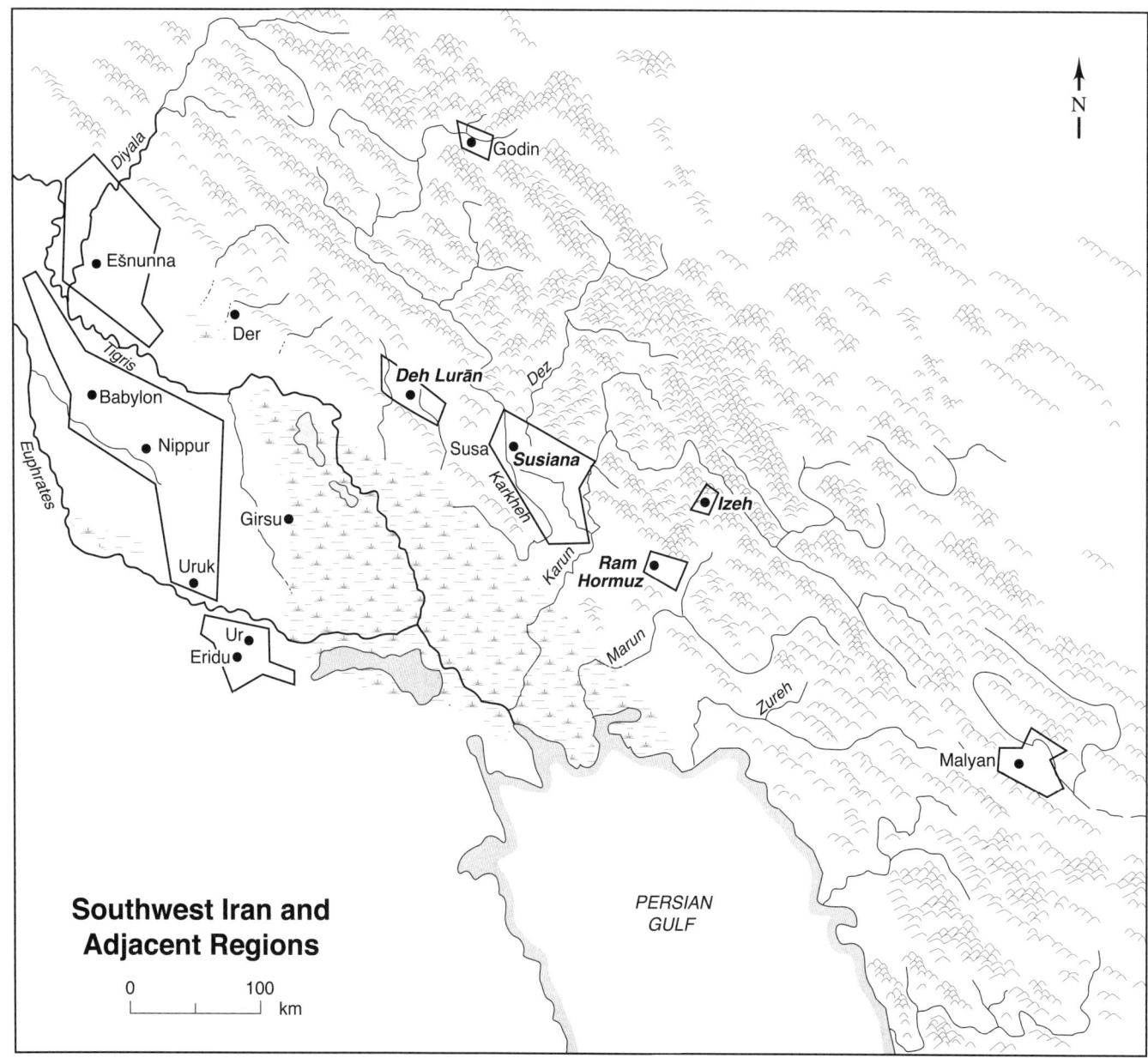

Figure 1.1. Greater Mesopotamia and the Deh Lurān Plain. (Names in bold designate survey areas. Other names are sites mentioned in the text.)

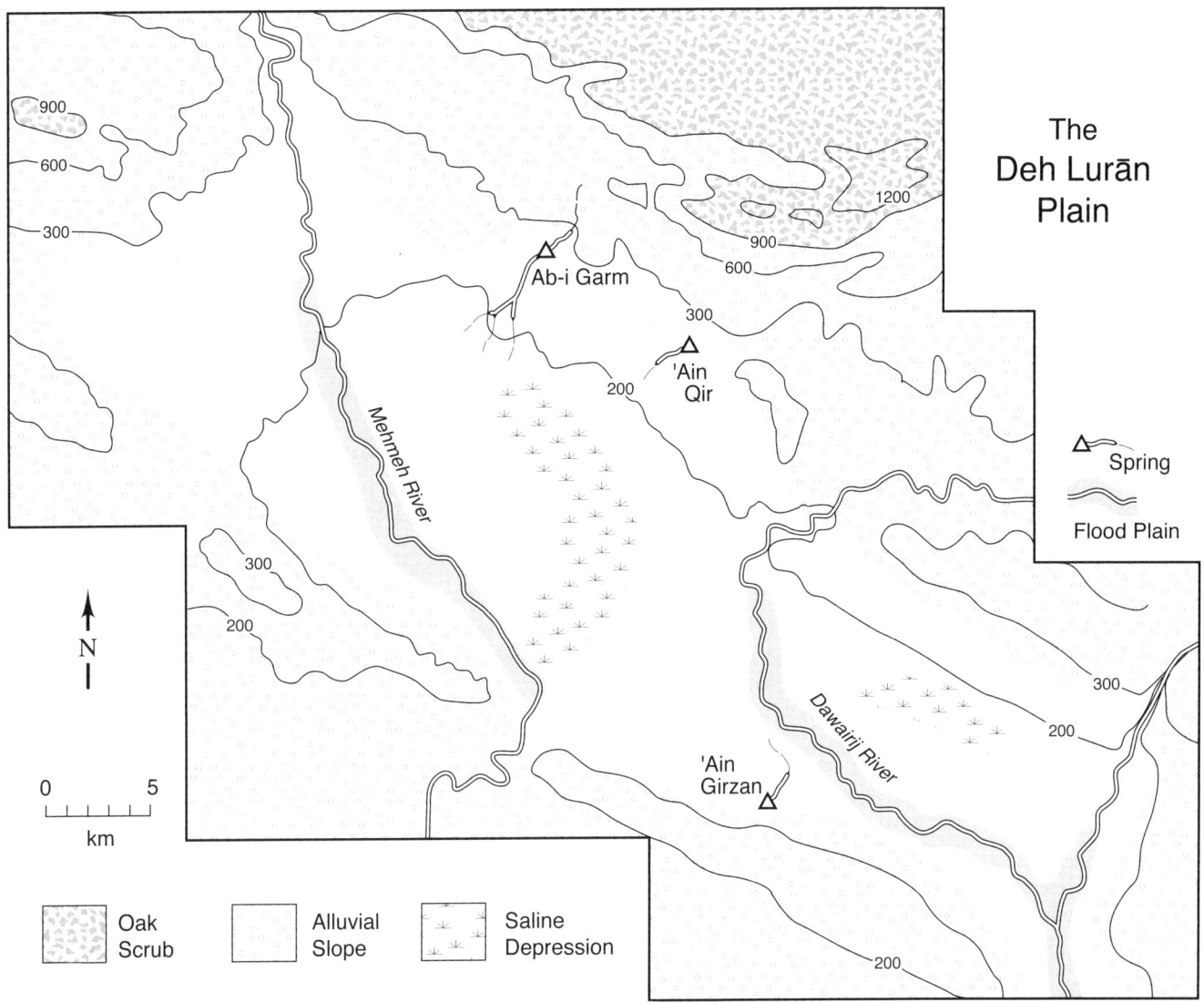

Figure 1.2. Natural environments of the Deh Lurān Plain.

Archéologique Française; and Henry Wright and Robert C. Gibbs from the University of Michigan. Elizabeth Carter, then of the Oriental Institute of the University of Chicago, participated in the survey and analyzed all Elamite materials. Carter's contributions are incorporated in the present study.

The methods used during the survey are outlined in Chapter 2. Sites found early in the main survey season were given temporary field numbers followed by the letter "x." Later, serial numbers were assigned in approximate geographic order from southeast to northwest. Since it had been learned that Braidwood had used only names, rather than numbers, for the sites he had visited, Neely began his renumbering with sites DL-1 to DL-19 in the northwest foothills and alluvial slopes of the plain near the town of Deh Lurān, passed over the already assigned numbers DL-20 to DL-33, scattered around the central plain, and continued with DL-34 in the southeast to DL-331 in the northwest. Thus, for example, DL 81x became DL-292. The latter number series has been used in this report, though future researchers may find that some of the sherds and field notes have "x" numbers, and for this reason we also give these numbers in Chapter 4. Most sherds from the 1968–1969 survey are at present curated at the Texas Archaeological Research Laboratory of the University of Texas at Balcones Research Center in Austin, Texas. They will be permanently housed in the Peabody Museum of Yale University, New Haven, Connecticut.

Portions of the survey data discussed in this volume have been cited in several previous manuscripts and publications, and since we have modified our interpretation of the data over the years, some discussion of these earlier publications is warranted. No statements regarding early historic settlement pattern were made as a result of the earlier surveys of 1961 to early 1968. Some results of the 1968–1969 survey were used in Carter's doctoral dissertation (Carter 1971), in an article on later irrigation systems (Neely 1974b), in the final report on the excavations at Tepe Farukhābād (Carter in Wright 1981), and in a general overview of early Iran (Carter and Stolper 1984). Since these publications, we have made several reassessments of the chronology and we have restudied all the sherds and changed the dates of occupation of some sites.

Since 1969, several archaeologists have visited the region, but the only formal survey undertaken was that conducted in 1976 in the area between the Susiana and Deh Lurān Plains by Pierre de Miroschedji of the Délégation Archéologique Française. Miroschedji (1981) focused on sites ruled by the Neo-Elamite, Achaemenid, and Seleucid dynasties during the Ist millennium B.C. This survey recorded a number of sites on the eastern end of the Deh Lurān Plain, and assigned site numbers beginning with the designation "DK." The valuable information provided by Miroschedji is cited in the site descriptions in Chapter 4, and in contributions to the synthetic discussion in Chapter 7.

We have knowledge of what has happened to archaeological sites on the Deh Lurān Plain since 1976 only through recently available satellite imagery. It is clear that some sites have been damaged or destroyed during the Iran-Iraq War, but more have been damaged by modern mechanized agriculture. Given such dangers as mines and explosives, it seems likely that further research will be limited for many years to come.

Plan of the Present Work

Following this introduction, we present in Chapter 2 the research design for the survey and the methods used both to recover the data demanded by the research design and to analyze these data in the laboratory. In Chapter 3, we undertake the task of describing those artifact categories of each phase, primarily ceramics, which were found on the surfaces of sites. A series of phase assemblages are illustrated in Figures 3.1 to 3.4. Counts of the occurrence of many of the categories of artifacts found on the sites are included in Appendix A tables. As the understanding of these materials improves, scholars may use these figures and counts to reinterpret the settlement data. In Chapter 4 the details of each site are presented. Included are: (1) the numerical designations for the site and the site's name, if known; (2) three grid locations in terms of longitude and latitude; of a Transverse Mercator Grid used on the maps of the National Iranian Oil Company, the only modern maps available to us; and of the Universal Transverse Mercator Grid, which is now replacing other schemes of geo-referencing; (3) the site's dimensions; (4) a brief statement about the site's environmental context; (5) a discussion of any features such as surface irregularities, stone constructions, unusual concentrations of artifacts, and later post-occupational disturbances; (6) discussion of the cultural phases present; (7) any comments necessary on the interpretation of the site's size or characteristics in particular phases; and finally (8) any published references. Maps of most of the sites, made by Neely with pace and compass, have been revised using "Google Earth" images. Detailed illustrations of many samples of sherds—with captions presenting category designations, detailed attribute observations, and references—are also included in this chapter. In Chapter 5, the evidence is synthesized in terms of successive settlement patterns. In Chapter 6, special issues of water management are discussed. Chapter 7 provides an overview of the documentary sources and integrates documentary and archaeological evidence. Finally, Chapter 8 presents some arguments about changing environments, economic processes, and political organization on the Deh Lurān Plain during the time of the early empires.

Acknowledgments

Our projects in Deh Luran would not have been possible without the support of the Director General of the Archaeological Service of the Ministry of Arts and Culture, Mr. 'Ali Pourmand, and the Director of the subsequently reorganized Iranian Centre for Archaeological Research, Dr. Firouz Bāgherzādeh. Our field representatives for survey were Mr. Sayid M. Khorramābādi in 1961 and Mr. M. Azimzādeh and architect Reza Zahedāni in 1969. We extend our sincere thanks to all our Iranian colleagues.

The authors and contributors wish to thank all the Deh Lurān project members who helped with the survey work, cited above. During the publication process, we were also helped by many others. We thank Frank Hole not only for his support in the field and his Preface, but for sharing his photographic archives. The team at eMap International, especially Sky Rubin and Justin Harmon, were indispensible in helping us get legitimate access to the Google Earth Imagery used on Plates 2 to 6. Kay Clahassey miraculously converted all our images and drawings into clear and expressive illustrations. Finally and most important, the editor of the Museum of Anthropology publications, Jill Rheinheimer, spent more than a year taking the manuscript to final publication with great skill and forbearance for the authors' scattered lives. We alone are responsible for any errors or deficits.

Chapter 2

Research Design and Methods

James A. Neely

In my segment of the 1968 proposal to the U.S. National Science Foundation I focused largely on an investigation of the irrigation systems. Three specific tasks were proposed:

(1) to undertake an intensive reconnaissance of the Deh Lurān Plain to discover and record the presence of water control and irrigation features;

(2) to conduct limited test excavations to help define the structure and function of these features as well as provide information for their dating; and

(3) to relate these data to the information that Anne and Michael Kirkby obtained on physiographic features and land use practices today so that a reasonable reconstruction of prehistoric conditions may be made.

Within the first weeks of fieldwork, however, the team became aware of two major problems that were impeding the work proposed: (1) it was evidently not going to be possible to determine the chronological placement of the vast majority of the water control and irrigation features and systems solely through a study of the features and systems themselves, and (2) a reasonable reconstruction of the prehistoric conditions would apparently not fully explain the structure and function of the features and systems without some knowledge of the prehistoric and early historic communities and the settlement patterns that developed during the long occupation of the Deh Lurān Plain. In order that the water control and irrigation features and systems might be seen in their proper chronological and socio-cultural perspective, we initiated a site survey. First we set out to record all of the sites on the central part of the Deh Lurān Plain. The number, complexity, and preservation of these sites, however, soon led us to realize that only a portion of the plain's area could be surveyed in the time available, and that a sample of the sites present would have to suffice. In response to this situation, we devised the "band" and "zone" sampling methodology (see below) by means of which approximately 70% of the plain was surveyed. We also worked with relatively high-altitude, low-resolution aerial photographs to plot the locations of the visible sites, as well as canals and other water control features, with as much precision as possible.

In the beginning of the field notebook, I wrote that it was the intent of the survey to examine an area of about one thousand square kilometers comprising most of the plain in order "to find, describe, and relate every site on the Deh Lurān Plain to the environment as well as to the society and culture of the prehistoric and historic inhabitants." In the preliminary report prepared in 1969 for the National Science Foundation (Hole ed. 1969), these interests were elaborated:

> The survey and test excavations were to accomplish the following objectives: 1) to discover and record the presence, temporal placement, structure, and function of prehistoric and early historic water control and irrigation systems; 2) to provide an overall picture of the relationship of such systems to the settlement patterns, cultivated fields, crop yields and population trends.

As a result of these interests, the survey team approached the denuded landscape of the Deh Lurān Plain with the intention of examining every square kilometer and mapping precisely all types of sites, occupational and otherwise, but we were limited by circumstances.

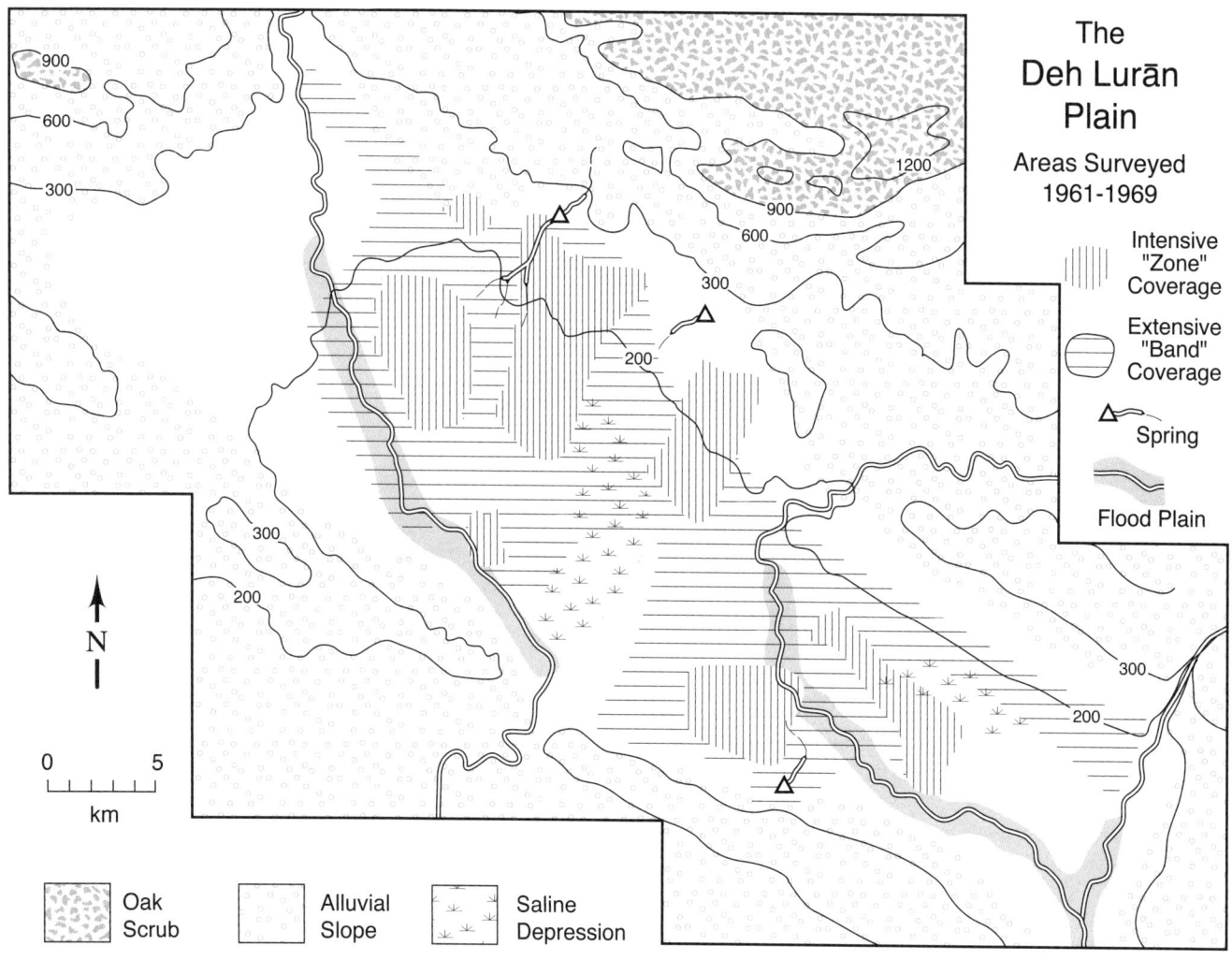

Figure 2.1. Survey coverage of the Deh Lurān Plain.

- We did not wish to survey only a sample of the area because we did not believe we could retrieve water control systems or patterns of emerging settlement hierarchies without full coverage (see Kowalewski and Fish 1990). However, as the season progressed, it became evident that because of the large number of sites, it would not be possible to complete the survey in the time available. Therefore, some portions of the survey area termed "zones" were fully surveyed. Other portions of the area were surveyed in a set of parallel "bands" or transects crossing the area (Fig. 2.1). It was hoped that some of the areas surveyed by the "band" technique could be fully surveyed with the "zone" technique in subsequent seasons, but this was never possible.

- We did not attempt to undertake a complete walking survey because the team was always small and because vehicle survey, giving the team the perspective of height, was usually possible. Only in certain areas, as noted below, did we attempt coverage on foot.

- We only rarely had a survey instrument that could be used to map each site precisely. Also, we did not have low-altitude, high-resolution air photographs that would have allowed detailed photogrammetric mapping. Usually, we were forced to use simpler techniques with Brunton pocket transit, tape, and pacing.

The field procedures on a typical day were as follows. The air photographs were studied the night before to locate major sites. The area selected for survey was visited early in the morning when the angle of the sunlight was low. The area was traversed by vehicle in parallel transects about 25 meters apart, looking for low mounds, stone footings, or sherd and lithic scatters. When a feature was found, it was located on a 1:50,000 aerial photograph in terms of nearby watercourses, roads, bushes, and larger mounds evident on the photos. Once a number of sites had been located or the angle of the light became higher, making it difficult to locate sites even at close range, the revisiting of sites was begun. At each, a pace and compass map was made and features were measured with a tape. The site was then divided into sectors, and diagnostic sherds and other artifacts were collected from each sector. In the evening, the locational information was copied onto base maps obtained from the National Iranian Oil Company. Sherds were washed and numbered during the next few days, and chronological assessments were made. Special circumstances required changes in this daily procedure. Survey of the rugged margins of the plain or areas with deep gulleys necessitated that the coverage be largely on foot. The study of large and topographically complex sites such as Tepe Musiyān (DL-20) required days of mapping and sectorial collecting in topographically defined areas. Large relatively featureless sites were divided into measured rectangular zones, conforming where possible to surface changes. On such sites, special attention was paid to erosional cuts and old excavations in order to assess stratigraphic accumulation. Sometimes, circumstances prevented recovery of large collections, and only a small type sample was retained for further study.

While the original 1961 survey was restricted to the central part of the plain between the Mehmeh and the Dawairij or Āb-Dānān Rivers, we intended the 1968–1969 survey to cover the entire plain, almost 1000 square kilometers. Though much more was examined than in 1961, it was not possible to cover the northwestern extremity of the plain, beyond a kilometer or so from the eastern bank of the Mehmeh. Also, coverage of the southeastern extremity, across a dangerous ford and distant from the project base in the village of Deh Lurān, was not as intensive as elsewhere. Fortunately, this is the area reexamined by Miroschedji (1981), so coverage is better than we had expected. All told, about 70% of the plain was examined, and within that part, we believe that about 80% of all of the visible sites extant on the plain were recorded.

At the end of the season, through the generosity of the Iranian Archaeological Service, all of the sherds collected during the survey were sent to the University of Texas at Austin for analysis. The procedures used were those used in analyzing ceramics from the excavated sites: an initial division into "wares" with characteristic clay bodies, inclusions, and colors; a further division into "types" based on surface treatment; and a final division into variants based on the vessel shape and decorative motifs (Hole, Flannery, and Neely 1969:109–11). Drawings of diagnostic ceramics from all sites were made. The ceramic variants in the samples in Texas and some of the samples at Yale have been counted, and the counts are presented in Appendix A. However, it is important to emphasize that these counts are not directly comparable to those given for the screened excavated samples in the various Deh Lurān excavation reports since plain body sherds were not uniformly recovered and retained for analysis. It is intended that the diagnostic sherds studied in Austin will be permanently curated at the Peabody Museum at Yale University in New Haven, Connecticut.

Though none of us have been able to return to the Deh Lurān Plain since 1976, we are fortunate to have recently gained access to "Digital Globe" imagery via the "Google Earth" web site. These images are in full color and have been ortho-rectified and with one exception we have been able to correct and improve our maps, to see recent damage to sites, and even to see architectural traces not evident in surface examinations. The exception is DL-54 on the east portion of the plain, which has been completely leveled for agriculture. We were, however, able to use our conventional 1961 aerial image to improve our sketch map of this site. We are deeply grateful to Google, to Digital Globe, and to eMap International for this unexpected chance to improve our observations.

Chapter 3

Ceramic Phase Indicators in Surface Assemblages

Elizabeth Carter and Henry T. Wright

The excavations at Tepe Farukhābād have resulted in the definition of a sequence divided into four successive cultural phases dating between the middle IIIrd millennium B.C. and the middle of the IInd millennium B.C. The presence of any of these phases on a site can be recognized if only a few diagnostic ceramics are present in a surface collection. Other phases, however, have been recognized only in surface collections, verified by comparison with nearby areas, but not yet identified in excavations at sites on the Deh Lurān Plain. The following discussion considers six of these phases. In this chapter, the assemblages are illustrated by complete or reconstructable examples of types common on the surface. Examples illustrated in the corpora of surface ceramics in Chapter 4 are cited in the figure captions.

The Later Early Dynastic Phase
(Figure 3.1)

Ceramics from the end of the Early Dynastic period are known from small excavated samples from Excavation A, Layers 1–5, and Excavation B, Layers 20–21, at Farukhābād (Wright 1981:168–73). Dittmann (1986) has shown that these layers date to the Early Dynastic III period in Mesopotamia. Most of these ceramics are made of a sandy fabric similar to that of the Early Dynastic I–II and earlier phases. The sherds, however, are thicker and the indications of fast wheel throwing are much more in evidence.

The characteristic surface assemblage of the later Early Dynastic phase on the Deh Lurān Plain has many conical cup and bowl fragments. The rims of conical cups (ED-RC-1: Fig. 3.1*a, c*) predominate. Straight-sided cups (ED-RC-3: Fig. 3.1*b*) are distinctive. The string-cut bases of conical cups (ED-BC-3) include shallow (Fig. 3.1*a*) and wide (Fig. 3.1*c*) variants. Round lip bowls (ED-RB-1) remain common. A heavier variant of the ledge rim bowl (ED-RB-3: Fig. 3.1*e*) develops. A carinated bowl (ED-RB-4, ED-RB-6: Fig. 3.1*d*) is added to the repertoire. Jar rims occur in flared round lip forms, both heavy (ED-RJ-1: Fig. 3.1*j*) and fine (ED-RJ-2) variants, as well as a flared flat lip variant (Fig. 3.1*f*). Heavier variants of both the ledge rim (ED-RJ-4: Fig. 3.1*l, o*) and band rim (ED-RJ-5: Fig. 3.1*m*) jars are common. Jar embellishments include the addition of heavy raised strips, either plain (ED-O-1: Fig. 3.1*j, o*) or impressed (ED-O-2: Fig. 3.1*l*). A new small carinated jar or goblet form (ED-RJ-6: Fig. 3.1*g, h*) was often painted in a variety of monochrome designs. Larger jars may have elaborate polychrome designs (ED-O-4, ED-O-6: Fig. 3.1*o*). Jars and bowls usually have flat bases (ED-B-2) or ring bases tooled on the wheel (ED-B-3), including a high variant (ED-B-4: Fig. 3.1*o*). Pinched or impressed ring bases (ED-B-5: Fig. 3.1*n*) are rare. The handmade gray ware jars of this phase, sometimes with crushed freshwater shell inclusions, are uncommon in surface assemblages (ED-RJ-7: Fig. 3.1*i*). Denticulate sickle blades are commonly associated with these ceramics.

The representations in the illustrations for Chapter 3 are of either complete forms known from the excavated sites or reconstructions based on sherds from survey sites illustrated in this volume. More detailed information will be found in the individual captions.

Key:

AM: Achaemenid   RB: rim of bowl
B: base   RC: rim of cup
BC: base of cup   RJ: rim of jar
BG: base of goblet SI: Šimaški
ED: Early Dynastic SJ: rim of straw-tempered jar
NE: Neo-Elamite   SM: Sukkalmaḫḫu-Middle Elamite
O: "other artifact"

Figure 3.1. Ceramic vessels of the Early Dynastic III phase.

a, Shallow conical cup (ED-RC-1 with ED-BC-3): Musiyān (DL-20) 5.2, Fig. 4.3a.

b, Straight-sided cup (ED-RC-3 with ED-BC-3): Musiyān (DL-20) 1.2, Fig. 4.3d; Aliabad (DL-71), Gautier and Lampre 1905: Fig. 268.

c, Wide conical cup (ED-RC-1 with ED-BC-3): Tepe Gārān (DL-34), Fig. 4.21b.

d, Carinated bowl (ED-RB-6): Musiyān (DL-20) 2.1, Fig. 4.4g.

e, Ledge rim bowl (ED-RB-3): Musiyān (DL-20) 6.2, Fig. 4.4i.

f, Flared flat lip jar (ED-RJ-3) with monochrome decoration (ED-O-3): Farukhābād (DL-32), Wright ed. 1981: Fig. 56m (B22).

g, Small carinated jar (ED-RJ-6) with monochrome decoration (ED-O-3): Tepe Sohz (DL-54), Neely and Wright 1994: Fig. IV.33s.

h, Small carinated jar or goblet (ED-RJ-6) with monochrome decoration (ED-O-3): Musiyān (DL-20) 2.3, Fig. 4.6m; Musiyān (DL-20) 3.2, Fig. 4.9b.

i, Gray ware jar with loop handles (ED-RJ-7): Farukhābād (DL-32), Wright ed. 1981: Fig. 66o (Exc. B24); Musiyān (DL-20) 1.1, Fig. 4.5a.

j, Flared round lip jar (ED-RJ-1) with raised strip (ED-O-1): Farukhābād (DL-32), Wright ed. 1981: Fig. 49b (Exc. A3).

k, Gray ware jar with flared round rim (ED-RJ-7): Musiyān (DL-20) 6.3W, Fig. 4.5d.

l, Ledge rim jar (ED-RJ-4) with impressed strip (ED-O-2): Musiyān (DL-20) 2.3, Fig. 4.6c, d; Tepe Sohz (DL-54), Neely and Wright 1994: Fig. IV.33p.

m, Band rim jar (ED-RJ-5): Musiyān (DL-20) 3.2, Fig. 4.6g.

n, Pinched ring base (ED-B-5): Farukhābād (DL-32), Wright ed. 1981: Fig. 56f–h (Exc. A1–3, B23).

o, Ledge rim jar (ED-RJ-4) with raised strip (ED-O-1), polychrome bands (ED-O-4), polychrome motifs (ED-O-6), and high ring base (ED-B-4): Aliabad (DL-71), Gautier and Lampre 1905: Fig. 286.

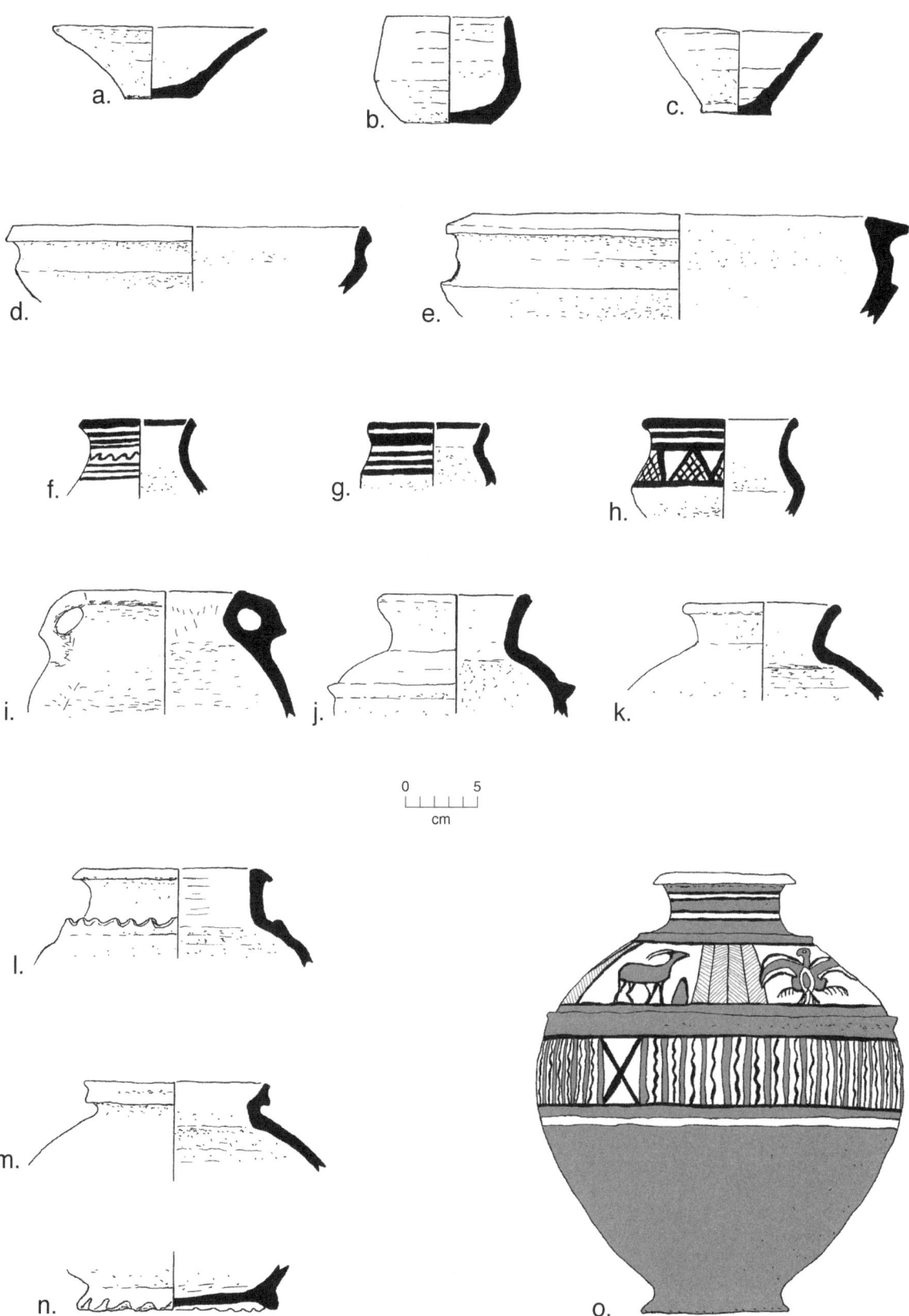

Figure 3.1. Ceramic vessels of the Early Dynastic III phase.

The Šimaški Phase
(Figure 3.2)

Ceramics of the Later Akkadian and Ur III period, termed the Šimaški period in southwest Iran (Carter 1971; Carter and Stolper 1984), are represented at Tepe Farukhābād in Excavation B, Layers 15–19, and at the great regional center of Susa in both the older Ville Royale B excavation of Roman Ghirshmann (Gasche 1973) and in the more recent Ville Royale I excavation of Carter (1980). We can thus discuss Šimaški phase ceramics with some confidence. The Farukhābād sample is an assemblage of broken domestic pottery recovered by screening a small exposure of the floors and the fills of modest mud brick houses in what was then a small village. It is a small sample, but useful because it is from the Deh Lurān Plain. The Ville Royale B sample is a selection of complete or near complete vessels from a large exposure of substantial urban housing in the south part of the Ville Royale quarter of Susa. The Ville Royale I sample is a complete sample from a small exposure of both fills and floors of substantial housing and from graves. While the Susa samples may be biased by cultural selection for use in graves and latrines, they are well dated by association with texts and sealings and are invaluable sources of information on complete vessel forms.

Some of these ceramics are made of a fabric with sandy inclusions similar to that of the Early Dynastic and earlier phases, but some, particularly the larger basins, have fabrics with a mixture of sand and vegetal material rare in the earlier phases. There is no evidence that the handmade gray ware continued in use on the Deh Lurān Plain.

The Šimaški phase is not represented by any single-period occupation site on the Deh Lurān Plain. Šimaški ceramics usually occur with later Elamite ceramics. Nonetheless, we can recognize Šimaški occupation because these ceramics have a distinctively different fabric and because several distinctive forms are known from excavations. The rims of wide conical cups (SI-RC-1: Fig. 3.2d), incurved cups (SI-RC-2: Fig. 3.2e), and distinctive indented band rim cups (SI-RC-3: Fig. 3.2f) are common on Deh Lurān sites, as they are in other regions. All appear to have string-cut bases (SI-BC-1). Round lip bowls (SI-RB-1) occur. A heavier variant of the ledge rim bowl, both ledge (SI-RB-2: Fig. 3.2g) and oblique rims (SI-RB-3: Fig. 3.2h), continues from the preceding phase. A large basin with a heavy grooved band rim (SI-RB-4: Fig. 3.2j), often with wavy incised decoration (SI-O-2), is distinctive to the Šimaški phase on the Deh Lurān Plain. A large incurved basin or vat with flat lip and heavy ribs (SI-RB-5: Fig. 3.2i) was added to the repertoire. Goblets and jars occur with flared rims and round lips (SI-RG-1: Fig. 3.2a; SI-RJ-1: Fig. 3.2b), as well as a flared flat lip variant (SI-RJ-2: Fig. 3.2c), often with a marked angle at the neck and shoulder junctures. A widely known distinctive jar with an undercut band rim and ribbed shoulders (SI-RJ-3, SI-O-1: Fig. 3.2k) occurs consistently. A jar variant with an indented band rim (SI-RJ-4: Fig. 3.2l) also occurs. A small heavy jar variant with flared neck and thick band rim (SI-RJ-5: Fig. 3.2m) or ledge rim (SI-RJ-6: not illustrated) is well dated at Nippur (McMahon 2006: Pl. 119) and may be part of the Deh Lurān Šimaški assemblage, though it was not excavated at Farukhābād. A widely known distinctive jar with a heavy grooved band rim (SI-RJ-6: Fig. 3.2n) occurs. Jars and bowls have flat or disc bases (SI-B-1: Fig. 3.2b, h) or high ring bases tooled on the wheel (SI-B-2: not illustrated). Wavy incising (SI-O-2: Fig. 3.2g, j) and combing (SI-O-3) are common on jars and basins.

Most Šimaški ceramics in Deh Lurān have close parallels in the central Susiana Plain, but there are also parallels in Lower Mesopotamia proper. Forms such as the indented band rim bowl (SI-RB-3), the jar with under-cut band rim (SI-RJ-3) and with ribbed shoulder (SI-O-1), and the jar with a grooved collar (SI-RJ-7) spread throughout the Mesopotamian world during the expansion of the Sargonid Dynasty and the IIIrd Dynasty of Ur. It is interesting that some of the decorative modes of these widespread and presumably prestigious forms are used on local ceramic variants, for example the grooved rim basin (SI-RB-4) not attested from adjacent regions.

The Sukkalmaḫḫu and Middle Elamite Phases
(Figure 3.3)

Ceramics equivalent to the Later Larsa and Old Babylonian periods in Lower Mesopotamia, termed the Sukkalmaḫḫu period in southwest Iran (Carter 1971), are represented at Tepe Farukhābād in Excavation B, Layers 14 to 11b, and at the great regional center of Susa in the Ville Royale A Excavation, Layers XV to XII (Gasche 1973), which has a wealth of complete vessels. Ceramics equivalent to the succeeding Cassite period, first termed the "Transitional Phase" by Carter (1971) and now called the Early Middle Elamite Phase, are represented at Tepe Farukhābād in Excavation B, Layers 11a–2 and at Susa in the Ville Royale A Excavation, Layers XV to XII (Gasche 1973). Surprisingly, modern publications on Later Middle Elamite (often called "classical") ceramics are few. Fortunately, we do have a publication on well-stratified material from Ville Royale II, Layers 13–10 (Miroschedji 1981:12–17, Figs. 10–16). We can thus discuss Sukkalmaḫḫu and Middle Elamite ceramics with confidence.

While there is much continuity in vessel forms, the change from Šimaški to Sukkalmaḫḫu ceramics involves a qualitative change in the preparation of potting clays. Most of the ceramics of these periods are made of a distinctive ware with vegetal temper, probably from animal dung, and a greenish, greenish-white, or white color. Earlier wares have a higher proportion of mineral inclusions. Why this seemingly sudden change occurred has rarely been discussed and never adequately explained. Sukkalmaḫḫu and Middle Elamite surface assemblages usually include characteristic goblet, jar, and basin sherds of this ware, with only some of the larger basins and jars having fabrics with sand or crushed sherd inclusions. Recovery of only a few sherds of these variants indicates site use during the early or middle IInd millennium, or both.

Sukkalmaḫḫu surface assemblages have many sherds from goblets with ovoid bodies and flared necks. These include rims (SM-RG-1: Fig. 3.3*a, c*), button bases (SM-BG-4: Fig. 3.3*a*), or small disc bases (SM-BG-1: Fig. 3.3*b*). Small thickened round lip bowls (SM-RB-1: Fig. 3.3*f*), out-turned round lip bowls (SM-RB-2: Fig. 3.3*g*), and deeper ledge rim bowls (SM-RB-3: Fig. 3.3*h*) occur, but only the last is common. A large slightly incurved basin with a flat lip and heavy ribs (SM-RB-4: Fig. 3.3*i*) and a large basin with a ledge rim (SM-RB-5: Fig. 3.3*j*) continue. A large basin with a square rim (SM-RB-6: Fig. 3.3*k*) is introduced. A large basin with a heavy band rim (SM-RB-7: Fig. 3.3*l*), several with impressed appliqué strips (SM-O-2), is distinctive. A jar with ledge rim (SM-RJ-1: Fig. 3.3*m*) is common, but a heavier square rim jar (SM-RJ-2: Fig. 3.3*n*) is not. A jar with flared oblique rim (SM-RJ-3: Fig. 3.3*p*) as well as a band rim jar (SM-RJ-4: Fig. 3.3*q*) are common. Jars and bowls have flat bases (SM-B-1) or high ring bases tooled on the wheel (SM-B-2: not illustrated). Horizontal grooves, wavy combing (SI-O-1), and impressed appliqué strips (SM-O-2: Fig. 3.3*l*) occur on jars and basins.

Early Middle Elamite surface assemblages on the Deh Lurān Plain are always mixed with Sukkalmaḫḫu assemblages. Our knowledge comes primarily from the small excavated Farukhābād sample. These ceramics are similar to Sukkalmaḫḫu assemblages, but can be recognized by several characteristic features. They too have many sherds from goblets with ovoid bodies and flared necks. Flared rims (SM-RG-1: Fig. 3.3*a, c*) continue, but a tall variant (SM-RG-2: Fig. 3.3*b*) develops. Button bases (SM-BG-4: Fig. 3.3*a*) and flat disc bases (SM-BG-1: Fig. 3.3*b*) continue, but a variant with a small ridge at the juncture of the body and base (SM-BG-2: Fig. 3.3*d*) becomes common. Variants with interior plugs (SM-BG-3: Fig. 3.3*e*) appear for the first time. Small thickened round lip bowls (SM-RB-1: Fig. 3.3*f*), and the deeper ledge rim bowl (SM-RB-3: Fig. 3.3*h*), continue, but the out-turned variant (SM-RB-2: Fig. 3.3*g*) does not. The large incurved basin with flat lip and heavy ribs (SM-RB-4: Fig. 3.3*i*) and the large basin with ledge rim (SM-RB-5: Fig. 3.3*j*) continue. The large basin with a square rim (SM-RB-6: Fig. 3.3*k*) and the large basin with a heavy band rim (SM-RB-7: Fig. 3.3*l*) are not definitely attested in Middle Elamite times. The jar with a ledge rim (SM-RJ-1: Fig. 3.3*m*), the square rim jars (SM-RJ-2: Fig. 3.3*n*), and the jars with a flared oblique rim (SM-RJ-3: Fig. 3.3*p*) continue. A flared concave or slightly indented variant of the band rim jar (SM-RJ-5: Fig. 3.3*o*), apparently occurring on narrow high-necked jars, appears. Jars and bowls have flat bases (SM-B-1) or high ring bases tooled on the wheel (SM-B-2: not illustrated).

Late Middle Elamite assemblages are not represented by an excavated sample from the Deh Lurān area, and we therefore must rely on the well-excavated sample from Susa Ville Royale II (Miroschedji 1981: Figs. 11–15). The vessel fabrics have vegetal inclusions and are predominantly brown rather than green. Goblets become long and narrow, with very high rims. The goblet bases with an interior plug become common (SM-BG-3: Fig. 3.3*e*), but the base variant with the ridge is not attested. The small, thickened round lip bowl and the ledge rim bowl continue, as do the large basins with flat or ledge rims. Jars, often with low necks, have thickened round rims, oblique rims, and concave or indented band rims (Miroschedji 1981: Figs. 24, 25, 36, 43, 44). Future study of local assemblages may show that these jars are good indicators of Later Middle Elamite occupation. Jars and bowls have flat bases or high ring bases as before. At present, it is difficult to recognize Later Middle Elamite ceramics in a collection with much Earlier Middle Elamite and Sukkalmaḫḫu Elamite material.

The Neo-Elamite and Achaemenid Phases
(Figure 3.4)

After the end of the Middle Elamite period at about 1100 B.C. there appears to have been little settlement either on the Deh Lurān Plain or in the surrounding areas for several centuries. Unfortunately we have no excavated Neo-Elamite or Achaemenid assemblages from the Deh Lurān Plain. The well-excavated assemblages from Ville Royale II, Levels 9–6 (Miroschedji 1981a), are our only point of reference. However, a few identifiable Neo-Elamite ceramics have been recognized from the surfaces of sites in the Deh Lurān area.

Excavated Neo-Elamite ceramics have vegetal inclusions in the tradition of Middle Elamite ceramics, but are usually light brown to reddish-brown in color and are rarely white or green.

The conical cup had a renaissance in Neo-Elamite times, if not before. The rims of conical cups, both straight (NE-RC-1) and slightly incurved (NE-RC-2: Fig. 3.4*a*), and their string-cut bases (NE-BC-1) are common at Susa, but few have been recognized in surface collections from the Deh Lurān area. Low wide mouthed goblets with a neck ridge (NE-RG-1) and a flat base (NE-BG-1) are common at Susa, and Miroschedji (1981b: Fig. 64:3) recovered one from Tepe Patak (DL-35) on the Deh Lurān Plain. Small, thickened round lip bowls (NE-RB-1), noted at Susa (Miroschedji 1981a: Fig. 20:8–10), were found at the same site (Miroschedji 1981b: Fig. 64:2). Ledge rim and oblique rim bowls (NE-RB-2) occur at Susa (Miroschedji 1981a: Figs. 19, 20:1–6, 10), but have not been recorded on the Deh Lurān Plain. Large basins with flat rims (NE-RB-3) continue from earlier Elamite times, and they maintain a light-brown or greenish body color (Miroschedji 1981a: Fig. 26), but none have been recognized from the Deh Lurān Plain. Jars with high flared necks and round or flat rims (NE-RJ-1) (Miroschedji 1981a: Figs. 23, 34:76–79, 35:12–15), often with pointed bases (NE-RJ-3), are common at Susa. Miroschedji (1981b: Fig. 64:4) recovered one from Tepe Patak (DL-35) on the Deh Lurān Plain. Jars with oblique or ledge rims (NE-RJ-2) are common at Susa (Miroschedji 1981b: Figs. 24:1, 12, 14, 25:5, 6) and a burnished example with neck ridge (Fig. 3.4*c*), duplicated at Susa (Miroschedji 1981b: Fig. 23:10), was found on the high mound at Tepe Gārān (DL-34). The flared concave or slightly indented variant of the band rim

Figure 3.2. Ceramic vessels of the Šimaški phase.

a, Goblet with flared round lip (SI-RG-1) with flat base (SI-B-1): Musiyān (DL-20) 3.1, Fig. 4.8a; Susa Ville Royale I3, Gasche 1973: Fig. 46.

b, Flared round lip jar (SI-RJ-1) with flat base (SI-B-1): Farukhābād (DL-32), Wright ed. 1981: Fig. 90g (B18).

c, Flared neck jar with oblique rim (SI-RJ-2) with flat base (SI-B-1): Musiyān (DL-20) 1.1, Fig. 4.10c; Susa Ville Royale I4, Gasche 1973: Fig. 42:3.

d, Wide conical cup (SI-RC-1) with flat base (SI-BC-1): Susa Ville Royale BVII–VI, Gasche 1973: Pl. 1:14–17.

e, Incurved cup (SI-RC-2) with flat base (SI-BC-1): Musiyān (DL-20) 3.2, Fig. 4.3e; Susa Ville Royale BVII–VI, Gasche 1973: Pl. 6:6–12.

f, Indented rim cup (SI-RC-3) with flat base (SI-BC-1): Musiyān (DL-20) 3.2, Fig. 4.3j–m; Susa Ville Royale I4, Gasche 1973: Fig. 41:1–4.

g, Ledge rim bowl (SI-RB-2) with wavy incising (SI-O-2): Musiyān (DL-20) 3.2, Fig. 4.11h.

h, Oblique rim bowl (SI-RB-3): Musiyān (DL-20) 3.2, Fig. 4.4j, k; Susa Ville Royale BV, Gasche 1973: Pl. 5:3, 4.

i, Flat lip ribbed basin (SI-RB-5): Musiyān (DL-20) 3.3, Fig. 4.11a.

j, Grooved band rim basin (SI-RB-4) with wavy incising (SI-O-2): Musiyān (DL-20) 3.2, Fig. 4.10g, h.

k, Band rim jar (SI-RJ-3) with ribbed shoulder (SI-O-1): Musiyān (DL-20) 3.1, Fig. 4.10a; Susa Ville Royale I8, Gasche 1973: Fig. 33:1; Nippur WFXIII–VII, McMahon 2006: Pl. 110.

l, Indented band rim jar (SI-RJ-4): Musiyān (DL-20) 3.2, Fig. 4.10b.

m, Heavy band rim jar (SI-RJ-5): Musiyān (DL-20) 3.2, Fig. 4.9c; Nippur WFIX–VIII, McMahon 2006: Pl. 119.

n, Grooved band rim jar (SI-RJ-6): Musiyān (DL-20) 5.1, Fig. 4.10f; Susa Ville Royale BVII, Gasche 1973: Pl. 16:11.

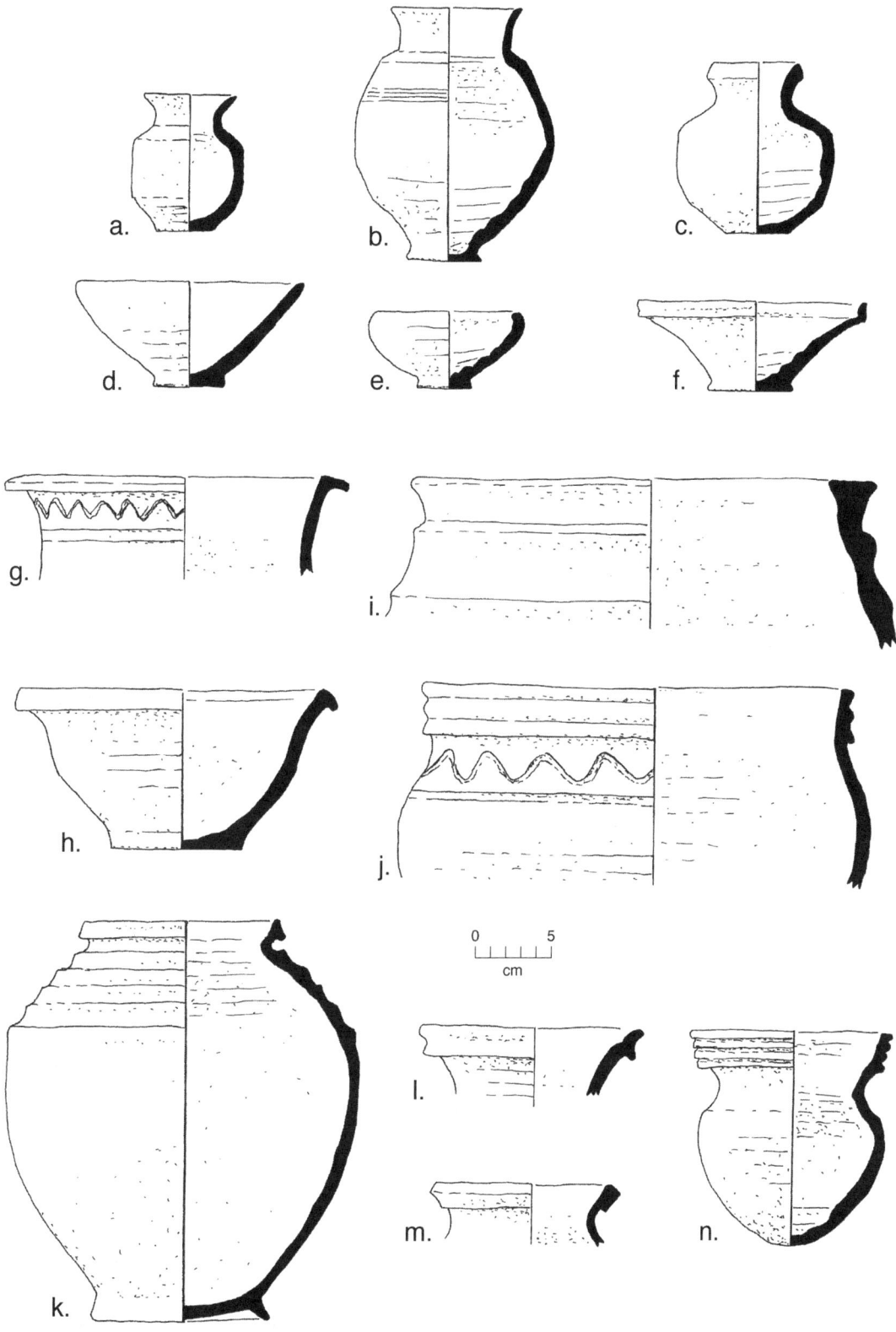

Figure 3.2. Ceramic vessels of the Šimaški phase.

Figure 3.3. Ceramic vessels of the Sukkalmaḫḫu and Middle Elamite phases.

a, Goblet with flared neck (SM-RG-1) with button base (SM-BG-4): Patak (DL-35)12, Fig. 4.26c; Susa Ville Royale AXI–XIII, Gasche 1973: Pl. 21:7, 9, 12, 20–23, 28, 29.

b, Tall goblet with high flared neck (SM-RG-2) and flat base (SM-BG-1): Gārān (DL-34) general collection, Fig. 4.22c.

c, Goblet with flared round lip (SM-RG-1) with flat base with inner plug (SM-BG-3): Farukhābād (DL-32), Wright ed. 1981: Fig. 90c (BF16).

d, Goblet base with ridge (SM-BG-2): Musiyān (DL-20) 5.1, Fig. 4.8d.

e, Small goblet base (SM-BG-3) with inner plug: Patak (DL-35) 12, Fig. 4.26d.

f, Thickened lip bowl (SM-RB-1): Tenel Ramon (DL-27) 2.1, Fig. 4.16n; Farukhābād (DL-32), Wright ed. 1981: Fig. 85c (B12); but see also Susa Ville Royale A, Gasche 1973: Pl. 19L.

g, Outcurved rim bowl (SM-RB-2): Tenel Ramon (DL-27) 1.4, Fig. 4.16p; Farukhābād (DL-32), Wright ed. 1981: Fig. 85b (B12); Suse Ville Royale AXIII–XV, Gasche 1973: Pl. 7, 14–16.

h, Ledge rim bowl (SM-RB-3): Gārān (DL-34) 1, Fig. 4.22i.

i, Flat rim basin (SM-RB-4): Tenel Ramon (DL-27) 2.1, Fig. 4.18e; Farukhābād (DL-32), Wright ed. 1981: Fig. 86n (B13).

j, Ledge rim basin (SM-RB-5): Patak (DL-35) 6–7, Fig. 4.27c; Farukhābād (DL-32), Wright ed. 1981: Fig. 86m (B11–13).

k, Square rim basin (SM-RB-6): Tenel Ramon (DL-27) 1.4, Fig. 4.18g; Farukhābād (DL-32), Wright ed. 1981: Fig. 86i (B13–16).

l, Band rim basin (SM-RB-7) with impressed strip (SM-O-2): Patak (DL-35) 6–7, Fig. 4.27b; Farukhābād (DL-32), Wright ed. 1981: Fig. 86L (B11–13).

m, Ledge rim jar (SM-RJ-1): Tenel Ramon (DL-27) 2.1, Fig. 4.17g; Farukhābād (DL-32), Wright ed. 1981: Fig. 85k (B12).

n, Square rim jar (SM-RJ-2): Patak (DL-35) 2, Fig. 4.26j; Farukhābād (DL-32), Wright ed. 1981: Fig. 85o (B12), m (B11–13).

o, Flared concave band rim jar (SM-RJ-5): Patak (DL-35) 7, Fig. 4.26l; Susa Ville Royale AXI–XI, Gasche 1973: Pl. 18:3, 4.

p, Oblique rim jar (SM-RJ-3): Patak (DL-35) 2, Fig. 4.26g.

q, Band rim jar (SM-RJ-4): Patak (DL-35) 6–7, Fig. 4.26k; Farukhābād (DL-32), Wright ed. 1981: Fig. 86b (B11–13).

Ceramic Phase Indicators in Surface Assemblages 19

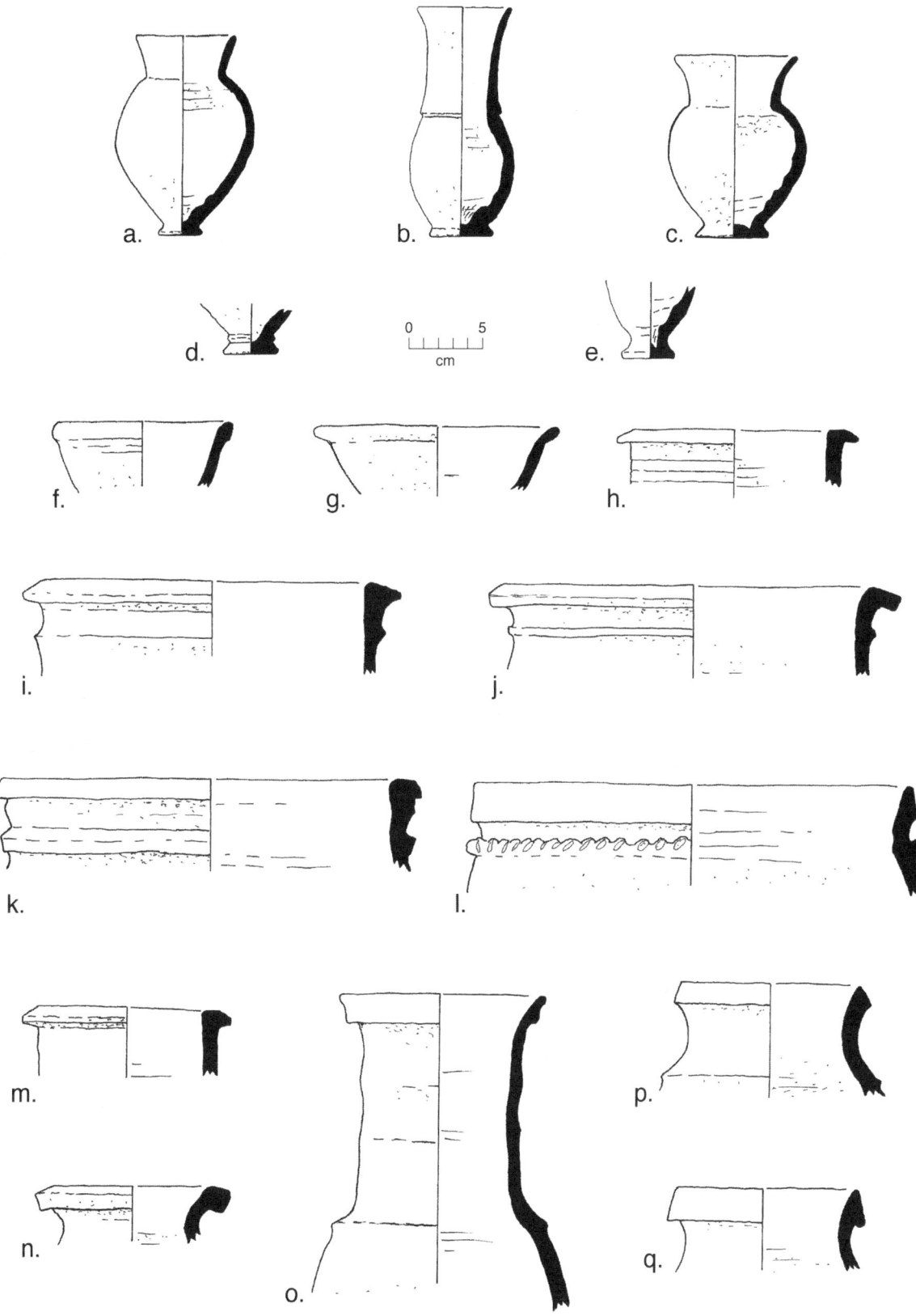

Figure 3.3. Ceramic vessels of the Sukkalmaḫḫu and Middle Elamite phases.

Figure 3.4. Ceramic vessels of the Neo-Elamite and Achaemenid phases.

a, Conical cup (NE-RC-2) with string-cut base (NE-BC-1): Susa Ville Royale II8–9, Miroschedji 1987: Figs. 17:6–11, 6:33:1–7.

b, Wide goblet with neck ridge (NE-RG-1) and flat base (NE-BG-1): Susa Ville Royale II8–9, Miroschedji 1987: Fig. 22:1–10.

c, Jar with beaded rim and neck ridge (NE-RJ-2): Gārān (DL-34) 1x, 4.24*l*; Susa Ville Royale II9, Miroschedji 1987: Fig. 23:10.

d, Conical cup (AM-RC-1) with string-cut base (AM-BC-1): Susa Ville Royale II5, Miroschedji 1987: Fig. 7:5, 6.

e, Band rim basin (AM-RB-6): Gārān (DL-34) '0' (north and west periphery), Fig. 4.23*a*; Susa Ville Royale II5, Miroschedji 1987: Fig. 11:4.

f, Fine carinated bowl (AM-RB-5): Susa Ville Royale II5, Miroschedji 1987: Fig. 7:7.

g, Glazed indented rim cup (AM-RC-3): Gārān (DL-34) general collection, Fig. 4.23*j*; Susa Ville Royale II4, Miroschedji 1987: Fig. 9:10.

h, Glazed base (AM-BC-2): Gārān (DL-34) general collection, Fig. 4.23*k*; Susa Ville Royale II4, Miroschedji 1987: Fig. 9:15.

i, Carinated bowl with ledge rim (AM-RB-4): Gārān (DL-34) 3, Fig. 4.23*i*; Susa Ville Royale II5, Miroschedji 1987: Fig. 7:12; Apadana 6B, Boucharlat 1987: Fig. 57:1.

j, Glazed ledge rim cup (AM-RC-2) with flat base (AM-BC-2): Susa Ville Royale II4, Miroschedji 1987: Fig. 9:6.

k, Carinated shallow bowl with beaded rim (AM-RB-2): Patak (DL-35) 2, Fig. 4.28*j*.

l, Carinated shallow bowl with beaded rim (AM-RB-2): Gārān (DL-34) 3, Fig. 4.23*e*; Susa Ville Royale II5, Miroschedji 1987: Fig. 9:1.

m, Carinated deep bowl with beaded rim (AM-RB-3): Gārān (DL-34) 3, Fig. 4.23*h*.

n, Carinated shallow bowl with beaded rim and painted bands (AM-RB-2): Gārān (DL-34) 6, Fig. 4.23*g*.

o, Neckless jar with beaded rim (AM-RJ-2): Gārān (DL-34) general collection, Fig. 4.24*d*.

p, Flared neck jar with beaded rim (AM-RJ-3): Gārān (DL-34) '0,' Fig. 4.24*h*.

q, Flared neck jar with beaded rim (AM-RJ-3): Gārān (DL-34) 1, Fig. 4.24*e*.

r, Straight neck jar with beaded rim (AM-RJ-4): Patak (DL-35) 13, Fig. 4.28*n*; Susa Ville Royale II4, Miroschedji 1987: Fig. 16:5.

s, Neckless jar with beaded rim (AM-RJ-2): Gārān (DL-34) 1, Fig. 4.24*d*; Susa Ville Royale II4, Miroschedji 1987: Fig. 15:5, 6.

t, Flared neck jar with beaded rim and neck ridge (AM-RJ-5): Patak (DL-35) 17, Fig. 4.28*o*; Susa Ville Royale II5, Miroschedji 1987: Fig. 16:3.

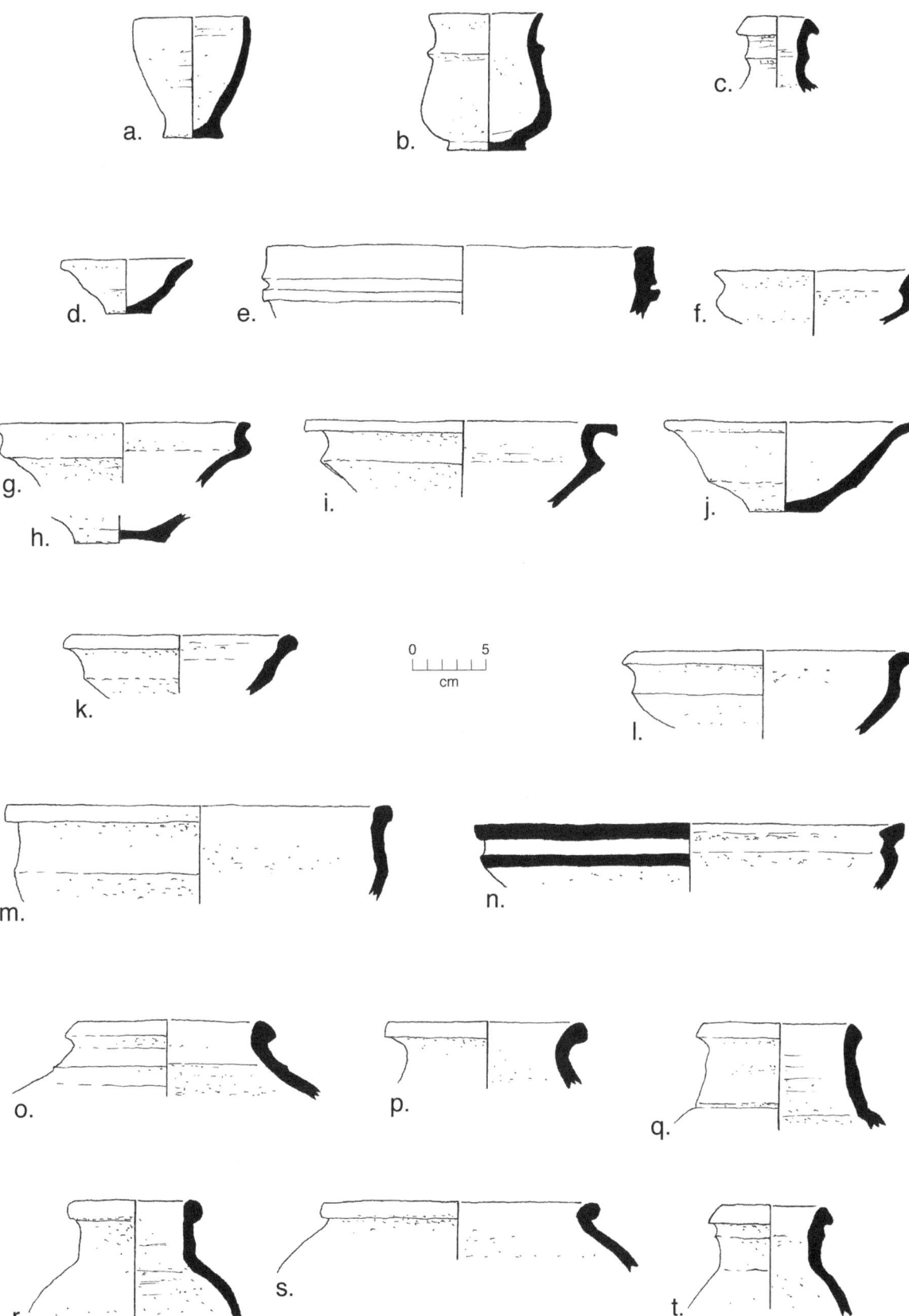

Figure 3.4. Ceramic vessels of the Neo-Elamite and Achaemenid phases.

(Miroschedji 1981b: Fig. 25:8–14) occurs at Susa on distinctive jars with low contracting necks (NE-RJ-3), but these have not been noted on Deh Lurān sites. Neo-Elamite jars and bowls usually have round or flat bases, but none have been recognized in surface collections.

The Achaemenid ceramics were first recognized in the Deh Lurān area because of their distinctive fabric and unmistakable large bowl forms (Miroschedji 1981b: Fig. 63). Assemblages excavated at Susa in stratigraphic context in the Ville Royale II, Levels 5–4 (Miroschedji 1987; Boucharlat 1987), though small, offer the best-dated parallels to the Deh Lurān occurrences. It is important to note, however, that these contexts are pits rather than floors, and they seem to be a later Achaemenid assemblage thought to date from the late sixth to late fourth centuries B.C.

The most recognizable Achaemenid vessels have a notably hard fabric with fine to medium sandy inclusions, pink or brown body colors, and light brown surfaces, smoothed or even burnished. Some larger jars (AM-RJ-2: Fig. 3.4o) and basins (AM-RB-6: Fig. 3.4e) also have vegetal inclusions, and small conical cups (AM-RC-1: Fig. 3.4d) have predominantly vegetal inclusions. The change to hard sandy fabrics must have occurred during early Achaemenid times, for which archaeologists have no well-excavated samples. There is no plausible explanation for this change.

The range of Achaemenid vessel parts we have recognized in the survey collections is limited. We have no doubt that when Achaemenid contexts are excavated in Deh Lurān, particularly when the wares other than the unmistakable hard sandy ware are studied, more will be identifiable. Cups include the small flat-based conical cup with vegetal inclusions noted above (AM-RC-1, AM-BC-1: Fig. 3.4d). This has not yet been recognized in the survey collections. Also noted are larger cups with a sandy fabric with a glaze now oxidized to a whitish powder, among which ledge rim (AM-RC-2: Fig. 3.4j) and indented rim (AM-RC-3: Fig. 3.4g) variants predominate. Both have tooled ring or flat bases (AM-BC-2: Fig. 3.4h, j). Simple open bowls with rounded or thickened lips (AM-RB-1) are known from Susa (Miroschedji 1987: Figs. 10:1, 12:1) but have not been recognized in the Deh Lurān collections. Bowls are predominantly shallow carinated forms with thickened beaded rims (AM-RB-2) ranging from medium-sized (Fig. 3.4k, l) to large (Fig. 3.4n), all of hard sandy ware. (The painted bands on this last example are unique, and may indicate a Hellenistic date.) A minority of the larger bowls are deeper, straight-sided carinated forms (AM-RB-3: Fig. 3.4m), also with thickened beaded rims. A few bowls of a burnished ware have traces of vegetal and calcite inclusions as well as sand, and have strongly carinated bodies and large ledge rims (AM-RB-4: Fig. 3.4i). A few bowls from Susa are of a fine ware without inclusions with a rounded carination and a flared rounded rim (AM-RB-5: Fig. 3.4f), but these have not been noted in the Deh Lurān collections. The only distinctive basin, known from Susa and from a single Deh Lurān example, is of a hard sandy ware; it has a prominent band rim and a ribbed body (AM-RB-6: Fig. 3.4e). Straight or flared neck round rim jars (AM-RJ-1) are known from Susa (Miroschedji 1987: Figs. 13:5, 14:3, 8), but have not been recognized in the Deh Lurān collections. Jars are predominantly a range of forms with beaded rims, and without necks (AM-RJ-2: Fig. 3.4o, s), with flared necks (AM-RJ-3: Fig. 3.4p, q), or with straight necks (AM-RJ-4: Fig. 3.4r), and sometimes with neck ridges (AM-RJ-5: Fig. 3.4t). Only flat bases are attested on the larger bowls and jars, but the sample is small.

A striking feature of the Deh Lurān Achaemenid ceramics is the strong representation of the beaded rim bowls. These are not common in the excavated samples from Susa. It seems likely that different activities required more large bowls in the Deh Lurān area. It is possible that these were economic activities, such as more processing of milk products. Alternatively, Deh Lurān domestic life may have emphasized social activities different from those of Susa involving different customs of food preparation and serving. Excavations of domestic contexts are needed.

Chapter 4

The Archaeological Sites and Their Interpretation

James A. Neely and Henry T. Wright

This chapter presents the recorded details of each site that produced artifacts dated from about 2550 B.C., the latest portion of the Early Dynastic period, until 325 B.C., the end of the Achaemenid period. We intentionally reconsider the Later Early Dynastic material discussed in the first survey volume (Neely and Wright 1994), in part because we can now better date material from previously discussed sites, and in part because we wish to integrate this material with what is known from the record of cuneiform texts and other documents.

The dating of the sites is based upon our present chronological understandings and utilizes the artifact samples conserved in collections at the Texas Archaeological Research Laboratory in Austin and at Yale University in New Haven. Ceramics illustrated by Gautier and Lampre (1905), in their still useful report on the project of the Délégation en Perse in the spring of 1903, and by Pierre de Miroschedji (1981), in his report of his survey under the auspices of the Délégation Archéologique Française en Iran in 1977, were also used. There are many additional questions about these sites that we could answer if it had been possible for us to return to the Deh Lurān Plain. Unfortunately, this has not been possible.

The following is a catalog listing only the sites having occupations dating between 2550 B.C. and 325 B.C. recorded on the Deh Lurān Plain. The sites are presented in the order of their designated survey numbers.

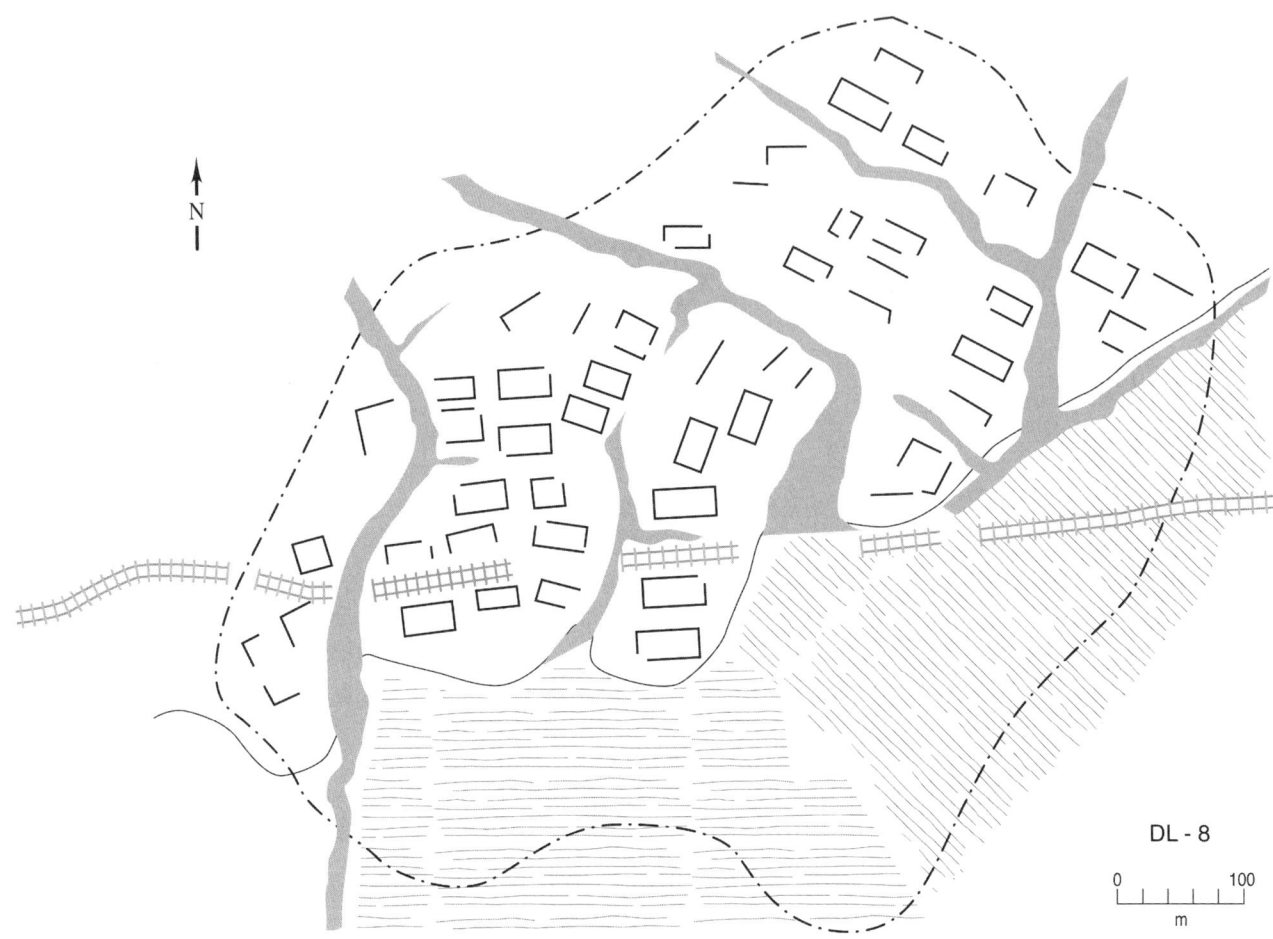

Figure 4.1. Map of Tāfuleh (DL-8). (Gray lines mark recent watercourses; parallel lines are fields.)

Number: DL-8 (DL-7X, 8X)
Name: Tāfuleh or Tawfuleh
Grid: E 47°12'23" / N 32°41'23"
  NIOC:[1] E 1706.4 – N 1189.1
  UTM:[2] E 707, 058 / N 3, 619, 108
Dimensions:
  Length: ca. 750 m (NE-SW)
  Width: ca. 550 m (NW-SW)
  Area: ca. 41.3 ha

Modern Environmental Context: At the lower edge of the northeastern piedmont slope of the plain near its juncture with the alluvial fan of the Mehmeh River about 450 m northeast of the present channel of the river.

Features: Tāfuleh is a low but relatively large site (Fig. 4.1) composed of many smaller mounds probably marking separate architectural complexes. We have prepared a plan of these features using measurements made from both older aerial images and recent satellite images. The southern part of the site has recently been leveled for agriculture, and the course of the narrow gauge railway to Amara—a jeep track at the time of our visit, but perhaps taking advantage of an earlier canal or road feature—is now barely visible (Plate 7 *upper*).

Phases: Unfortunately, our sample of ceramics from Tāfuleh is small. Among a series of later mostly Seleucid sherds, to be discussed in our third and final volume, are several possible Achaemenid or earlier sherds. There is one rim of a larger ledge rim basin (cf. SM-RB-5), and one rim of a jar with a hard sandy body and a beaded rim (AM-RJ-3) oxidized to a reddish tan color. The former has a long history, but the latter is almost certainly Achaemenid. We hope to illustrate these few sherds in our final volume.

Comment: We suspect that this is a larger Hellenistic or later settlement on top of a relatively small Achaemenid site, but until further ground studies are done, we cannot estimate its size.

---

[1] National Iranian Oil Company.
[2] Universal Transverse Mercator.

Number: DL-20
Name: Tepe Musiyān (Mussian, Moussian)
Grid: E 47°21'02" / N 32°33'13"
   NIOC: E 1721.0 – N 1174.2 (center)
   UTM: E 720, 704 / N 3, 604, 238
Dimensions:
   Diameter: 550 m (N-S)
   Width: 310 m (E-W)
   Height: 17.5 m
   Circumference: 1475 m
   Area: 15.5 ha

Modern Environmental Context: On the alluvial fan of the Dawairij or Āb-Dānān River, about one km west of the present river channel.

Features: Tepe Musiyān is the largest single mound on the Deh Lurān Plain (Fig. 4.2, Plate 2). Many of its surface details must result from major IInd millennium B.C. and later construction on the site, but its basic configuration probably results from architectural accumulation during and before the IIIrd millennium B.C., discussed in our previous publication (Neely and Wright 1994:57–66). Here, we present the details of the mound's surface. We have used the topographic map of Gautier and Lampre (1905: Fig. 95), even though their contour lines do not agree precisely with their elevation points, to determine the dimensions of the site and to augment our description. Their plan actually differs only slightly from their written description of length and width (Gautier and Lampre 1905:62). Although the edges of the mound are quite irregular, the overall configuration in plan is rectangular. The southern half of the site is oriented north-south, while the northern half is slightly angled to about 15° east-of-north. The French expedition of 1903 worked for three months, undertaking five large excavations (labeled A to E on Fig. 4.2) and several smaller test units.

A thorough, albeit rapid, surface investigation of this site required six full days of fieldwork. This study, augmented by a study of the profile of Gautier and Lampre's "Excavation E" in 1963 (Hole, Flannery, and Neely 1969:65–72) and Joseph Caldwell's profiling of the large erosional cut in the northwestern portion of the site, has indicated occupation of Tepe Musiyān from at least the Mohammed Jaffar phase (ca. 7000 B.C.) through the Early Islamic period (ca. A.D. 1250).

Because of the complexities of this huge site, we will discuss its occupational history in necessarily general terms. The focus here is on the uppermost 8.0 to 9.5 m of accumulation during the Early Dynastic and later phases.

Many cobble wall foundations or footings can be seen eroding from the site. In contrast to earlier wall footings discussed in our previous monograph (Neely and Wright 1994:57), those noted in the northeastern quadrant of Musiyān were more massive in both the stone material used and the overall size of the structures. One measured structure was 15 by 4 m. We do not know the date of these footings. We also made observations on the large mud brick buildings partially exposed in Unit A in the north-central part of the site by Gautier and Lampre (1905:67–68, Fig. 97), probably of the latest IIIrd and early IInd millennium. Baked bricks, probably from IInd millennium phases, are widely scattered on the site.

The configuration of the mound and traces on the Digital Globe imagery suggest that at various points, probably late in its history, Musiyān was surrounded by large walls. An entryway is clearly visible on the west. Whether the foundations of these walls date to the last time the site was a major settlement, the Sukkalmaḫḫu Elamite phase of the early IInd millennium B.C., or earlier, to the Šimaški or Early Dynastic period or before, can be resolved only by excavation. However, it is likely that the walls were still maintained during the Elamite phases.

Phases: The earlier occupations of Musiyān are detailed in our previous study (Neely and Wright 1994). We commence here with a recapitulation of our observations on the remains of the earlier IIIrd millennium B.C. since no sherds of these phases were illustrated in the earlier volume, and since several additional samples of Musiyān sherds have been found in the collections since completion of the previous study in 1991.

Ceramics typical of the Early Dynastic I–II phase are common in our massive samples. Solid footed conical cups (ED-BC-1: Fig. 4.3*g*, *h*), narrow conical cups (ED-BC-2: Fig. 4.3*f*, *i*), high, fine, round lip jars (ED-RJ-2: Fig. 4.5*g*), fine expanded or ledge rim jars (ED-RJ-4: Figs. 4.5*i*, *j*, 4.9*e*), often painted with characteristic polychrome jar shoulder decoration (ED-O-6: Fig. 4.6*k*, *l*), and handmade jars of calcite tempered gray ware (ED-RJ-7: Fig. 4.5*b–d*) are all common in excavated contexts of this period at Tepe Farukhābād from layers 6–12 in Excavation A and layers 24–21 in Excavation B (Wright ed. 1981:95–135, 172–73). Mapping of goblet base and polychrome fragments in the original study (Neely and Wright 1994: Fig. IV.15 *lower right*) shows that occupation covered the entire 15-ha mound, and we would not be surprised to find that this central town was walled at this time.

The ceramics we term "Later Early Dynastic" or "Early Dynastic III" phase are defined based on samples from layers 1–5 in Excavation A and layers 20–21 in Excavation B from Farukhābād (Wright ed. 1981:95–135, 173–74). They may continue into subsequent earlier Akkadian times, but without careful stratigraphic excavations at Musiyān, we cannot easily differentiate what Carter (1971) tentatively termed the "Awan Phase" in Deh Lurān surface collections. The shallow, wide conical cups (ED-BC-3: Fig. 4.3*a*, *b*), heavier round lip jars (ED-RJ-1: Fig. 4.5*f*), heavier ledge rim jars (ED-RJ-4: Fig. 4.6*a*, *b*), sometimes painted with monochrome jar shoulder decoration (ED-O-5: Fig. 4.5*k*), and band rim jars (ED-RJ-5: Fig. 4.6*i*) had long histories, but were particularly common in this time span. Handmade jars of shell tempered gray ware (ED-RJ-7: Fig. 4.5*a*), typical of this period at Farukhābād; carinated goblets, often with monochrome decoration (ED-O-5: Fig. 4.6*m*), typical of this period at Susa (Carter 1980:23, Figs. 28, 29; Stève and Gasche 1971); and fenestrated stands widely reported in lower Mesopotamia, particularly in the Diyala area (ED-O-7: Fig. 4.7a–c), are all diagnostic. The mapping of carinated goblets with monochrome decoration and fenestrated stands (Fig. 4.12 *upper left*) indicates occupation covered all but the far south end of the mound. Given the widespread occurrence of conflict in this period (see Chapter 7), we suspect that the town was walled at this time.

Elements of the Šimaški phase, as defined through study of samples from layers 18–16 in Excavation B at Farukhābād (Carter 1981:200–204), as well as from excavations in Ville Royale BVI–V (Gasche 1973) and Ville Royale I3–4 (Carter 1980) at Susa, are common in all of our Musiyān samples. The rims of wide conical cups (ED-RC-1: Fig. 4.3a, b) continue. Incurved cups (SI-RC-2: Fig. 4.3e) and distinctive indented band rim cups (SI-RC-3: Fig. 4.3j–n) are common. Complete examples often have string-cut bases with a small ridge (SI-BC-1: Fig. 4.3l). Heavier variants of the ledge rim bowl, with both flat (SI-RB-2: Fig. 4.11e, h) and oblique (SI-RB-3: Fig. 4.4j, k) rims, continued. A large basin with a heavy grooved band rim (SI-RB-4: Fig. 4.10g, h), often with wavy incised decoration (SI-O-2), is distinctive to the Šimaški phase on the Deh Lurān Plain. A large incurved basin with flat lip and heavy ribs (SI-RB-5: Fig. 4.11a) was added to the repertoire. Goblets and jars occur with flared rims and round lips (SI-RG-1: Fig. 4.8a) as well as a flared neck, oblique lip variant (SI-RJ-2: Fig. 4.10c). A widely known distinctive jar with undercut band rim and ribbed shoulders (SI-RJ-3, SI-O-1: Fig. 4.10a) is common at Musiyān. A jar variant with indented band rim (SI-RJ-4: Fig. 4.10b) occurs. A small heavy jar variant with flared neck and thick band (SI-RJ-5: Fig. 4.9c) or ledge (SI-RJ-6: Fig. 4.9d) rim dated to Ur III times at Nippur (McMahon 2006: Pl. 119) is part of the Deh Lurān Šimaški assemblage. A widely known distinctive jar with heavy grooved band rim (SI-RJ-6: Fig. 4.10f) occurs. Jars and bowls have flat disc bases (SI-B-1: Fig. 4.8g, h) or high ring bases tooled on the wheel (SI-B-2: Fig. 4.8j). Shallow wavy incising (SI-O-2: Figs. 4.10g, h, 4.11b, e, f, h) and combing (SI-O-3) are common on jars and basins. The indented rim bowl, grooved rim jars and basins and jars with ridged shoulders are found in all parts of Musiyān (Fig. 4.12 *upper right*), indicating that occupation covered the entire mound, and we would not be surprised to find that this central town was walled at this time.

Sukkalmaḫḫu Elamite and earlier Middle Elamite ceramics share many types. These are locally well attested in samples from Layers 14 to 1 in Excavation B at Farukhābād (Carter 1981:204–9) and are common in our Musiyān samples. The surface assemblages have many sherds from goblets with ovoid bodies and flared necks. These include rims (SM-RG-1: Fig. 4.8b) and small disc or button bases (SM-BG-1: Fig. 4.8f). A deeper ledge rim bowl (SM-RB-3: Fig. 3.3h) is common. A large incurved basin with flat lip and heavy ribs (SM-RB-4: Fig. 3.3k) and a large basin with ledge rim (SM-RB-5: Fig. 4.11b) continue. A jar with ledge rim (SM-RJ-1: Fig. 4.9g) is common, and a heavier square rim jar (SM-RJ-2: Fig. 4.11g) occurs. Jars and bowls have flat bases (SM-B-1: Fig. 4.8g) or high ring bases tooled on the wheel (SM-B-3: Fig. 4.8k). Horizontal grooves, wavy combing (SM-O-1), and impressed appliqué strips (SM-O-2) occur on jars and basins.

The local Sukkalmaḫḫu Elamite assemblage is recorded from Layers 14 to 11 Lower in Excavation B at Farukhābād (Carter 1981:206–9), and from excavations in Ville Royale AXIV–XV (Gasche 1973). Typical of this phase are an out-turned round lip bowl (SM-RB-2), large basins with square rims (SM-RB-6: Fig. 4.11c) and band rims (SM-RB-7), and a jar with flared neck and oblique lip (SM-RJ-3: Fig. 4.9h). Characteristically, Sukkalmaḫḫu oblique lip jars and large basins with square rims, usually in the distinctive greenish vegetally tempered ware, are found in all parts of Musiyān (Fig. 4.12 *lower left*), indicating that occupation covered the entire mound. We would not be surprised to find that the town was walled at this time.

A local earlier Middle Elamite or Transitional Elamite assemblage was recorded from Layers 11 Upper to 1 in Excavation B at Farukhābād (Carter 1981), which has parallels in excavations in Ville Royale AXII–XI at Susa (Gasche 1973). Many forms continue from Sukkalmaḫḫu times, but a distinctive flared concave or slightly indented variant of the band rim jar (SM-RJ-5, not illustrated from Musiyān), apparently occurring on narrow high-necked jars, was introduced. Also, the button bases of goblets were frequently embellished with a raised strip at the juncture of body and base (SM-BG-2: Fig. 4.8d). The distinctive band rim jar and button base with ridge are widely scattered in all parts of Musiyān (Fig. 4.12 *lower right*). Neither variant is common. Their scarcity at Musiyān could be because these later layers have been badly eroded or because actual occupation was limited. It seems unlikely that a frontier town like Musiyān, even one diminished in size, lacked defenses, but we have no evidence of them.

Comment: The phase identifications and areas assigned to each phase above are tentative because of the problem of the transport of construction fill from place to place on the site. However, examination of the distribution maps (Fig. 4.12) shows the broad patterns of use. Earlier Early Dynastic I–II is found throughout the 15-ha town, but Later Early Dynastic III material is rare in the far south, suggesting an occupied area of 13 ha. Šimaški phase ceramics—dating from the later years of the Sargonid Dynasty and the full span of the Third Dynasty of Ur, both of which probably controlled Musiyān (see Chapter 7) into the Isin-Larsa period—are found throughout the 15-ha town site. Ceramics of the Sukkalmaḫḫu phase, when Elamite rulers probably controlled the town, are similarly dense and widespread, but there is a diminution of the area of ceramic scatter in succeeding earlier Middle Elamite times to less than 10 ha, with subsequent abandonment.

References: Carter 1971:226–30; Gautier and Lampre 1905:62–72; Neely and Wright 1994:57–67.

The Archaeological Sites and Their Interpretation 27

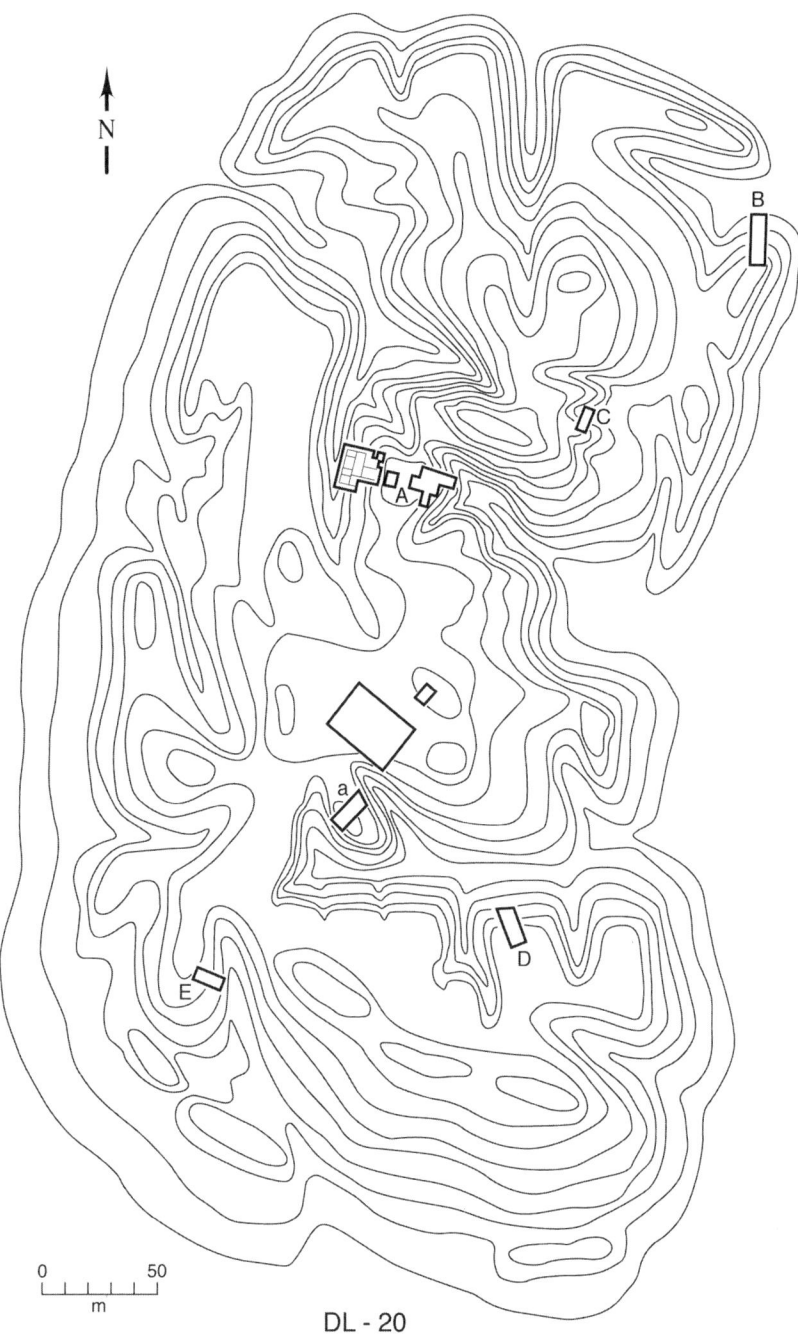

Figure 4.2. Map of Musiyān (DL-20).

Key:

AM: Achaemenid
B: base
BC: base of cup
BG: base of goblet
Dm: rim diameter
ED: Early Dynastic
NE: Neo-Elamite
NT: neck thickness
O: "other artifact"

RB: rim of bowl
RC: rim of cup
RHt: rim height
RJ: rim of jar
RT: rim thickness
SI: Šimaški
SJ: rim of straw-tempered jar
SM: Sukkalmaḫḫu–Middle Elamite
ST: side thickness

References to comparanda are as follows:

Farukhābād: Wright ed. 1981
Suse Ville Royale A and B: Gasche 1973
Suse Ville Royale I: Carter 1980
Suse Ville Royale II, Layers 8–6: Miroschedji 1981a
Suse Ville Royale II, Layers 5–1: Miroschedji 1987
Suse Apadana: Boucharlat 1987

Figure 4.3. Cups and bowls of the IIIrd millennium B.C. from Musiyān (DL-20).

a, Shallow conical cup (ED-RC-1 and ED-BC-3) (Area 5.2). Trace fine sand, calcite, and vegetal inclusions, Dm 14.5, height 5.1, base Dm 4.6, RT 0.53, ST 0.57, very pale brown (10YR 7/4) body (Farukhābād A7 [EDI]: Fig. 45*j*; Suse Ville Royale I, Level 7–8 [IVB]: Figs. 33:4, 34:2–3).

b, Shallow conical cup (ED-RC-1) (Area 6.3c). 5% fine sand inclusions, Dm 14, RT 0.43, ST 0.62, light red (2.5YR 6/5) body, very pale brown (10YR 8/3) surface (same comparanda as *a*).

c, Small conical cup (ED-RC-1) (Area 3.2). 5% fine sand inclusions, oval rim 10.0 × 9.2, height 6.1, oval base 5.5 × 4.5, RT 0.55, ST 0.83, pink (7.5YR 8/4) body (string cutting was too high, so hole was closed with a small patch in base; base subsequently heavily worn, breaking through the patch).

d, Straight-sided carinated conical cup (ED-RC-3) (Area 1.2). 5% fine sand inclusions, Dm 11, RT 0.38, ST 0.40, very pale brown (10YR 8/4) surface (Farukhābād B24 [Jemdet Nasr-EDII]: Fig. 46*b, c*).

e, Incurved conical cup (SI-RC-2 and SI-BC-1) (Area 6.5w). 5% calcite, fine sand, and vegetal inclusions, Dm 12, height 5.6, base Dm 4.0, RT 0.72, ST 0.72, light red (2.5YR 6/5) body (Suse Ville Royale BVII–VI [Šimaški] Group 5a: Pl. 6:8–12; Suse Ville Royale I, Level 4 [VB]: Fig. 41:1, Level 3 [VB]: Fig. 45:1).

f, Narrow constricted conical cup base (ED-BC-2) (Area 6.2x). Trace fine sand and vegetal inclusions, oval base 3.2 × 3.0, ST 0.83, pinkish gray (7.5YR 7/3) body (cf. Farukhābād B23 [EDI]: Fig. 45*d*).

g, Solid footed conical cup base (ED-BC-1) (Area 2.1). 5% fine sand inclusions, base broken, ST 0.51, very pale brown (10YR 7/3) body (Farukhābād B26 [Jemdet Nasr-EDI]: Fig. 45*g*).

h, Solid footed conical cup base (ED-BC-1) (Area 6.4e). Trace fine sand inclusions, base Dm 2.4, pink (7.5YR 7/4) body.

i, Narrow cup base (ED-BC-2) (Area 5.1). Trace fine sand inclusions, oval base 3.35 × 3.14, ST 0.77, greenish white (5Y 8/2) body (side and basal edge shaved) (Farukhābād B23 [EDI]: Fig. 45*d*).

j, Indented band rim bowl (SI-RC-3) (Area 2.3). Trace fine sand inclusions, Dm 18, RT 0.60, ST 0.60, RHt 2.20, pink (5YR 7/4) body, very pale brown (10YR 8/3) surface (Farukhābād B18 [Šimaški]: Fig. 84*a*; Suse Ville Royale BVII–VI [Šimaški] Group 6a: Pl. 7:4, 5, 11–19; Suse Ville Royale I, Level 4 [VB]: Fig. 41:2–4, Level 3 [VB]: Fig. 45:4).

k, Indented band rim bowl (SI-RC-3) (Area 5.1). Trace fine sand and calcite inclusions, Dm 18, RT 0.47, ST 0.62, RHt 1.30, pink (7.5YR 7/4) body (same comparanda as *j*).

l, Indented band rim bowl (SI-RC-3 and SI-BC-1) (Area 3.2). 5% fine calcite and fine sand inclusions, Dm 18, height 6.0, base Dm 6.2, RT 0.52, ST 0.65, RHt 1.17, very pale brown (10YR 7/4) body (same comparanda as *j*).

m, Indented band rim bowl (SI-RC-3) (Area 5.1). 5% fine sand inclusions, Dm 17, RT 0.50, ST 0.73, RHt ca. 1.73, pinkish gray (7.5YR 7/3) body (same comparanda as *j*).

n, Indented band rim bowl (SI-RC-3) (1968 General Collection). 5% fine sand inclusions, Dm 14, RT 0.65, ST 0.53, RHt 0.72, light red (7.5YR 7/3) body, weak red (10YR 5/4) slip (Farukhābād B16–17 [Šimaški]: Fig. 84*b*; Suse Ville Royale BVII [Šimaški] Group 6c: Pl. 7:15, 21, 22, 25–27; Ville Royale I, Level 4 [VB]: Fig. 41:1, Level 3 [VB]: Fig. 45:5).

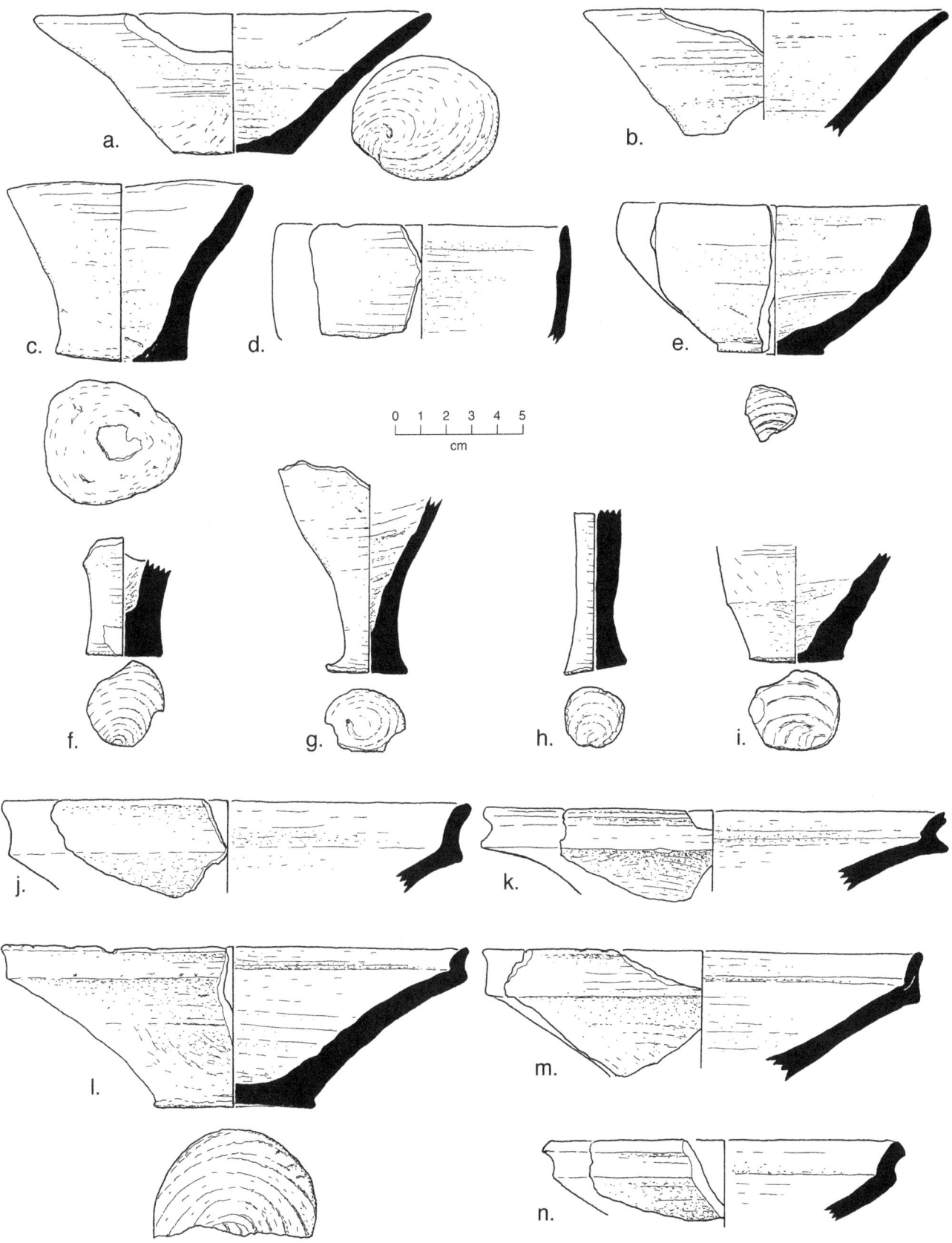

Figure 4.3. Cups and bowls of the IIIrd millennium B.C. from Musiyān (DL-20).

Figure 4.4. Bowls of the IIIrd millennium B.C. from Musiyān (DL-20).

a, Wide conical bowl (SI-RB-1) (Area 6.3w). 5% fine sand inclusions, Dm 24, RT 0.74, ST 0.72, pink (7.5YR 7/4) body, very pale brown (10YR 8/3) surface (Farukhābād B24 [Jemdet Nasr]: Fig. 46g; Suse Ville Royale I, Level 17 [IIIB]: Fig. 9:1).

b, Wide conical bowl (ED-RB-1) (Area 6.1x). 5% fine sand inclusions, Dm 18, RT 0.75, ST 0.83, pink (7.5YR 7/4) body, pink (7.5YR 7/3) surface (Farukhābād B22 [EDI]: Fig. 46d, B32 [Middle Uruk]: Fig. 46e; Suse Ville Royale I, Level 18 [IIIB]: Fig. 9:2).

c, Beveled lip bowl (ED-RB-2) (Area 6.3c). Trace fine sand inclusions, Dm 26, RT 1.57, ST 0.64, RHt 1.64, white (2.5Y 8/2) body (Farukhābād B31 [Late Uruk]: Fig. 47d, B25 [Jemdet Nasr]: Fig. 47j, A20 [Late Uruk]: Fig. 47L).

d, Beveled lip bowl (ED-RB-2) (Area 5.1). 10% medium sand inclusions, Dm 17, RT 1.02, ST 0.63, RHt 1.13, pink (7.5YR 7/4) body (same comparanda as c).

e, Beveled lip bowl (ED-RB-2) (Area 2.3). 10% fine sand inclusions, Dm 35, RT 1.80, ST 0.85, RHt 1.89, color not recorded (same comparanda as c).

f, Band rim bowl (Area 4.1). 5% medium sand inclusions, Dm 27, RT 0.97, ST 0.67, RHt 1.26, pink (5YR 7/3) body, reddish yellow (5YR 7/4) surface.

g, Carinated band rim bowl (ED-RB-6) (Area 2.1). 10% medium sand inclusions, Dm 27, RT 1.03, ST 0.55, RHt 1.22, very pale brown (10YR 7/3) body (Suse Ville Royale I, Level 17 [IIIB]: Fig. 12:6).

h, Ledge rim bowl (ED-RB-3) (Area 2.1). 10% fine sand inclusions, Dm ca. 24, RT 1.89, ST 0.88, RHt 1.39, very pale brown (10YR 8/3) body (Farukhābād A20 [Late Uruk]: Fig. 47e, A17 [Jemdet Nasr]: Fig. 47f).

i, Ledge rim bowl (ED-RB-3) (Area 6.2x). 10% fine sand inclusions, Dm ca. 40, RT 2.43, ST 1.16, RHt 1.72, white (10YR 8/2) body (same comparanda as h).

j, Oblique ledge rim bowl (SI-RB-3) (Area 3.2). 5% medium sand inclusions, Dm 24, RT 1.67, ST 0.80, RHt 1.30, reddish yellow (5YR 7/5) body (Farukhābād A6 [EDI]: Fig. 47h; Suse Ville Royale BV [Šimaški]: Pl. 5:3–4).

k, Oblique ledge rim bowl (SI-RB-3) (Area 5.1). 5% fine sand inclusions, Dm 39, RT 1.42, ST 0.87, RHt 1.17, pale yellow (5Y 8/3) body.

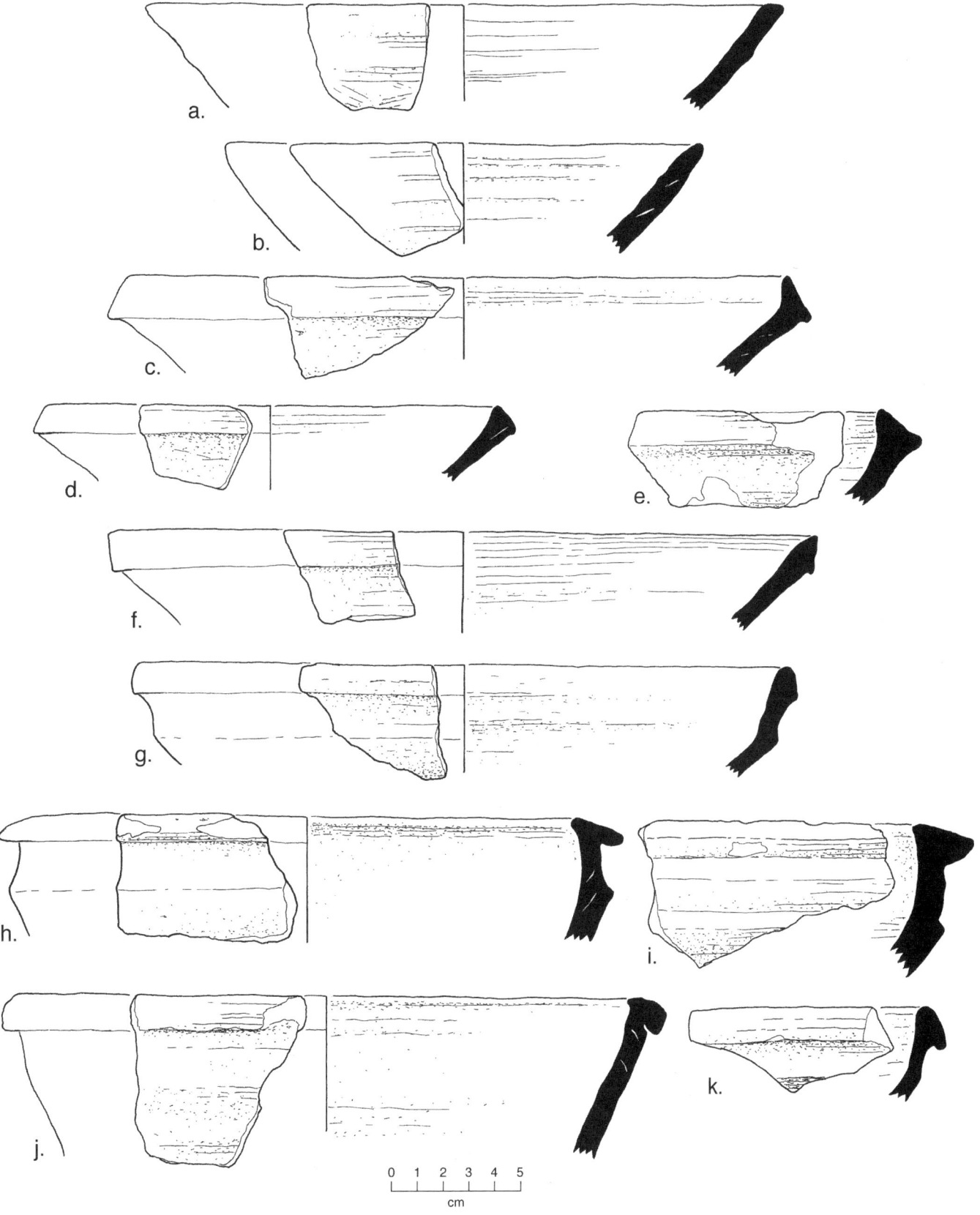

Figure 4.4. Bowls of the IIIrd millennium B.C. from Musiyān (DL-20).

Figure 4.5. Jar sherds of the IIIrd millennium B.C. from Musiyān (DL-20).

a, Gray ware jar with flared round rim (ED-RJ-7) (Area 1.1). 20% shell and calcite inclusions, Dm 14, RT 0.92, NT 0.72, ST 0.58, pale brown (10YR 6/3) body (Farukhābād B24 [Jemdet Nasr]: Fig. 66o, B27 [Jemdet Nasr]: Fig. 66q).

b, Gray ware jar with flared round rim (ED-RJ-7) (Area 5.1). 30% calcite inclusions, Dm 15, RT 0.50, NT 1.01, dark gray (N 4) body (Farukhābād: Fig. 66c).

c, Gray ware jar with flared round rim (ED-RJ-7) (Area 2.3). 30% calcite inclusions, Dm 18, RT 0.76, NT 0.74, ST 0.82, light brown (7.5YR 6/4) body (Farukhābād A16 [Jemdet Nasr]: Fig. 66f).

d, Gray ware jar with flared round rim (ED-RJ-7) (Area 6.3W). 20% calcite and vegetal inclusions, Dm 19, RT 0.93, NT 0.93, ST 0.90, dark gray (10YR 4/1) body (Farukhābād A6 [EDI]: Fig. 65b).

e, Round lip jar (ED-RJ-1) (Area 2.3). 5% fine sand inclusions, Dm 11, RT 0.71, NT 0.65, light red (10YR 6/6) body, very pale brown (10YR 8/3) surface (Farukhābād A7 [EDI]: Fig. 49d; Suse Ville Royale I, Level 18 [IIIB]: Fig. 13:8).

f, Round lip jar (ED-RJ-1) (Area 2.4). 5% fine sand inclusions, Dm 13, RT 0.71, NT 0.70, ST 0.69, very pale yellow (5Y 8/3) body (Farukhābād A5 [EDII+]: Fig. 49L).

g, Fine round lip jar (ED-RJ-2) (Area 5.1). 15% medium sand inclusions, Dm 12, RT 0.77, NT 0.55, white (5Y 8/2) body (Farukhābād B23 [EDII]: Fig. 49h, B21 [EDII+]: Fig. 49a).

h, Round lip jar (ED-RJ-1) (Area 5.1). Trace medium sand inclusions, Dm 10, RT 0.65, NT 0.66, very pale brown (10YR 8/3) body, red (10R 5/5) sludge (Farukhābād A3 [EDII+]: Fig. 49b).

i, Ledge rim jar (ED-RJ-4) (Area 5.1). 10% medium sand inclusions, Dm 13.5, RT 1.25, NT 0.90, RHt 1.13, white (2.5Y 8/2) body (Farukhābād B24 [Jemdet Nasr]: Fig. 53n, B22 [EDI]: Fig. 53q; Suse Ville Royale BV–VII [Šimaški] Group 29b: Pl. 36:1–3, 4, 7, 9, 10).

j, Ledge rim jar (ED-RJ-4) (Area 4.2). 5% medium sand and calcite inclusions, Dm 13, RT 0.94, NT 0.44, reddish-yellow (5YR 7/5) body, trace red paint (Farukhābād B24 [Jemdet Nasr]: Fig. 53n, A1–2 [EDII+]: Fig. 53m; Suse Ville Royale BV–VII [Šimaški] Group 29b: Pl. 36:1–3, 4, 7, 9, 10; Suse Ville Royale I, Level 18 [IIIB]: Fig. 13:7).

k, Flared expanded rim jar (ED-RJ-3) (Area 6.3c). 15% fine sand inclusions, Dm 10, RT ca. 0.96, NT 0.60, RHt ca. 0.70, trace of dark gray (N 4) painted neck band, trace of gypsum plaster plug in neck.

l, Conical spout (ED-O-7) (Area 1.9). 10% fine sand inclusions, reddish-yellow (5YR 6/6) body, very pale brown (10YR 8/4) surface.

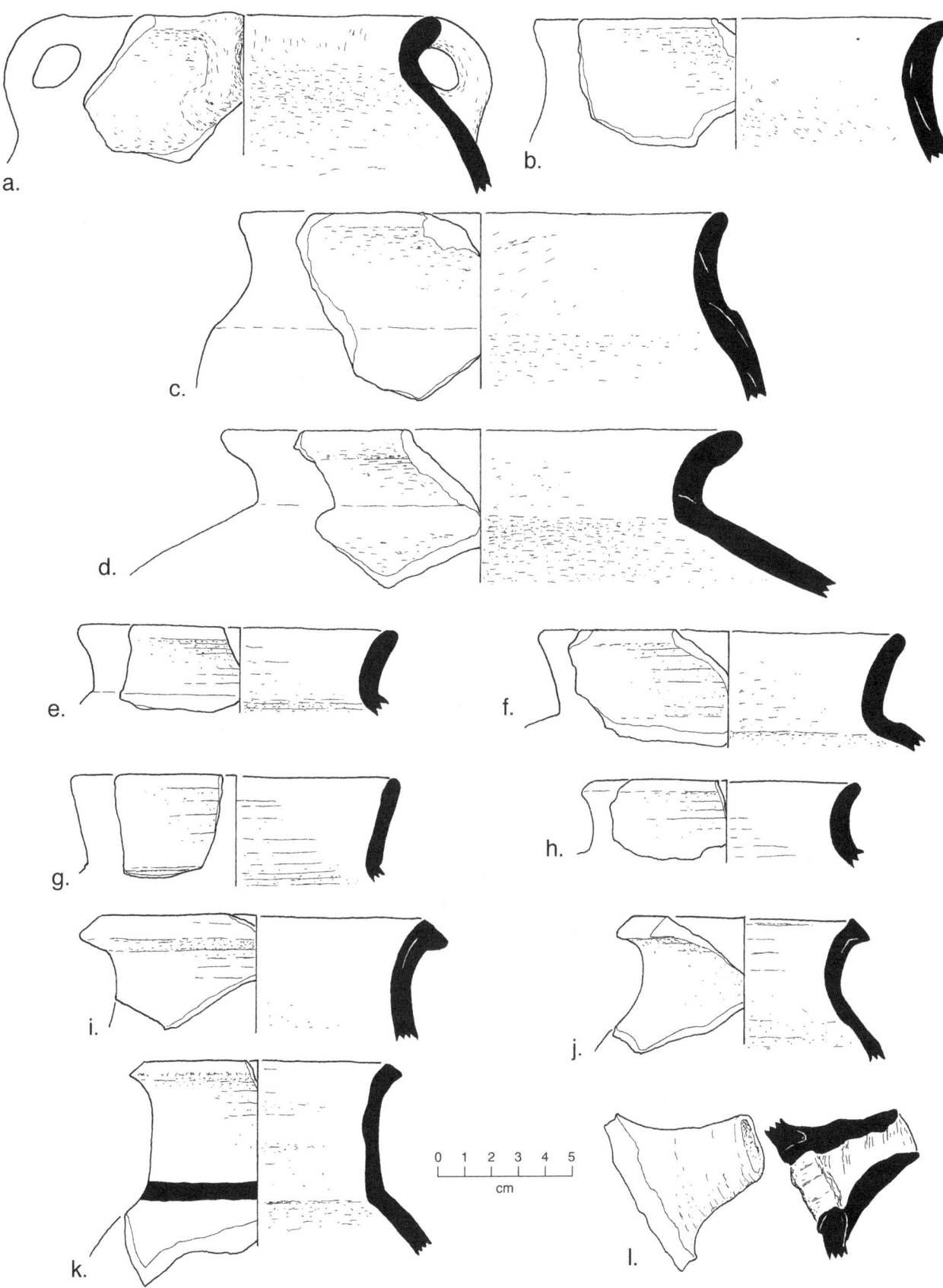

Figure 4.5. Jar sherds of the IIIrd millennium B.C. from Musiyān (DL-20).

Figure 4.6. Jar sherds of the IIIrd millennium B.C. from Musiyān (DL-20).

a, Ledge rim jar (ED-RJ-4) (Area 6.3cW). 10% fine sand inclusions, Dm 20, RT 1.68, NT 0.87, RHt 1.40, pink (5YR 7/4) body, white (10YR 8/2) surface, trace of red (10R 4/6) paint (Farukhābād A13 [Jemdet Nasr]: Fig. 53*g*).

b, Ledge rim jar (ED-RJ-4) (Area 5.1). 10% medium sand inclusions, Dm 15, RT 1.60, NT 0.84, RHt 1.01, white (2.5Y 8/1) body, yellow (10YR 7/5) sludge (Farukhābād A13 [Jemdet Nasr]: Fig. 53*g*, B20 [EDII+]: Fig. 58*a*).

c, Ledge rim jar (ED-RJ-4 with ED-O-2) (Area 2.3). 10% medium sand inclusions, Dm 16, RT 1.40, NT 0.62, RHt 0.79, white (2.5Y 8/2) body (Suse Ville Royale I, Level 9 [IVA]: Fig. 27:2).

d, Ledge rim jar (ED-RJ-4 with ED-O-2) (Area 6.1X). 5% medium sand inclusions, Dm 15, RT 1.99, NT 0.78, ST 0.92, RHt 1.28, white (2.5Y 8/2) body (note interior finger impressions) (same comparanda as *c*).

e, Band rim jar (ED-RJ-5) (Area 3.2). 10% fine sand inclusions, Dm 15, RT 1.03, NT 0.75, ST 0.85, RHt 1.44, very pale brown (10YR 8/3) body, white (10YR 8/2) surface (Farukhābād B20 [EDII+]: Fig. 54*g*; Suse Ville Royale BVI [Šimaški] Group 24b: Pl. 27:8).

f, Band rim jar (ED-RJ-5) (Area 2.1). 5% fine sand inclusions, Dm 13, RT 0.68, NT 0.52, ST 0.56, RHt 1.45, pink (7.5YR 7/4) body, white (2.5Y 8/2) surface (Farukhābād B23 [EDI]: Fig. 54*f*, B20 [EDII+]: Fig. 54*g*; Suse Ville Royale BVI [Šimaški] Group 24b: Pl. 27:8; Suse Ville Royale I, Level 6 [VA]: Fig. 39:11).

g, Band rim jar (ED-RJ-5) (Area 3.2). 10% calcite and fine sand inclusions, Dm 13, RT 1.03, NT 0.70, ST 0.80, RHt 1.43, very pale brown (10YR 8/3) body (Farukhābād B20 [EDII+]: Fig. 54*g*).

h, Band rim jar (ED-RJ-5) (Area 5.1). 10% fine sand inclusions, Dm 12, RT 0.80, NT 0.59, RHt 1.30, pink (5YR 7/4) body, very pale brown (10YR 8/3) surface. Suse Ville Royale BVI–VII [Šimaški] Group 24b: Pl. 27:9–10).

i, Band rim jar (ED-RJ-5) (Area 6.4E). 5% fine sand inclusions, Dm 15, RT 0.88, NT 0.66, ST 0.68, RHt 1.97, white (2.5Y 8/2) surface.

j, Jar shoulder with impressed strip (ED-O-2) (Area 3.2). 5% fine sand inclusions, body Dm 30, ST 0.62, strip thickness 1.41, imprint period 2.60, white (2.5Y 8/2) body.

k, Jar shoulder with polychrome decoration (ED-O-6) (Area 2). 5% medium sand inclusions, ST 0.71, white (10YR 8/2) body, weak red (10R 4/4) paint, black (10YR 3/2) paint (Farukhābād A10 [EDI]: Fig. 60*f*, A16 [Jemdet Nasr]: Fig. 60*d*, B22–23 [EDI]: Fig. 60*a*, *c*, *e*; cf. Suse Ville Royale I, Level 18B [IIIB]: Fig. 14:15).

l, Jar shoulder with polychrome decoration (ED-O-6) (Area 2.1). 10% medium sand, calcite, and vegetal inclusions, neck Dm ca. 12, NT 0.75, ST 0.84, white (2.5Y 8/2) body, red (10R 6/4) paint, black paint faint.

m, Goblet shoulder with monochrome decoration (ED-O-5) (Area 2.3). 5% medium sand and calcite inclusions, body Dm 13, ST 0.51, pale yellow (5Y 8/3) body, black (5Y 3/2) paint (Suse Ville Royale I, Level 9 [IVA]: Fig. 29:4).

n, Small ring base (ED-B-3) with red band (Area 4.3). 10% coarse sand inclusions, base Dm 6.5, ST 0.49, colors not recorded.

o, Pinched ring base (ED-B-5) (Area 5.1). 5% medium sand inclusions, base Dm 9.5, ST 0.81, white (2.5Y 8/1) body (Farukhābād A1–2 [EDII+]: Fig. 56*h*).

p, High ring base (ED-B-4) (Area 5.1). 10% medium sand and vegetal inclusions, base Dm 13.8, ST 1.24, very pale brown (10YR 8/3) body (battered and reworked, perhaps for use as lid) (Farukhābād A3 [EDII+]: Fig. 56*c*).

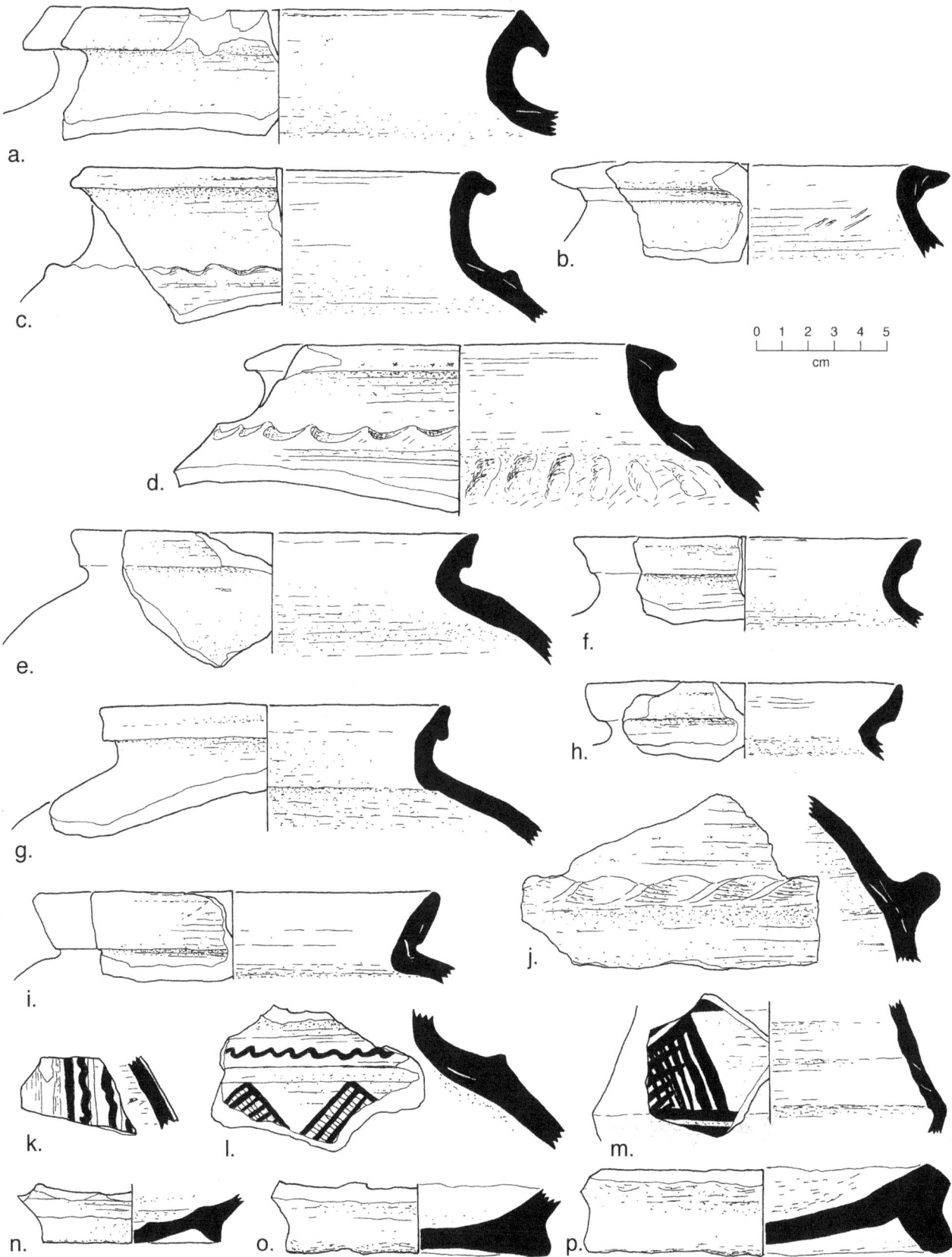

Figure 4.6. Jar sherds of the IIIrd millennium B.C. from Musiyān (DL-20).

Figure 4.7. Varia from Musiyān (DL-20).

a, Stand (ED-O-7) (Area 3.2). 15% medium sand inclusions, top Dm 14.5, base Dm 17, height 18.5, top RT 1.49, ST 0.94, top RHt 1.58, pale brown (10YR 7/4) body (Gautier and Lampre 1905: Fig. 279 [probably from Aliabad, DL-71, a cemetery for Musiyān]).

b, Stand (ED-O-7) (Area 6.2x). 10% fine sand inclusions, ST 1.32, very pale brown (10YR 8/3) body (orientation uncertain).

c, Stand (ED-O-7) (Area 5.1). 10% fine sand inclusions, base Dm 23, lower RT 1.45, ST 0.95, lower RHt 1.75, very pale brown (10YR 8/3) body.

d, Comb incised upright handle (Area 2/4). 10% fine sand inclusions, thickness 1.36, body color not recorded (cf. Khafajah [EDIII] Delougaz 1952: Pl. 86, Pl. 181).

e, Upright handle (Area 6.4E). 10% calcite, medium sand, and vegetal inclusions, pinkish gray (7.5YR 7/2) body (reverse shown; obverse damaged, destroying any possible decoration).

f, Horizontal lug (Area 6.4E). 5% fine sand inclusions, ST 0.69, light red (2.5YR 6/5) body, light gray (2.5Y 7/2) surface, trace of weak red (10R 5/4) paint.

g, Abstract figurine (Area 2.3). 10% fine sand inclusions, body thickness 2.07, pale yellow (2.5Y 8/3) body.

h, Female figurine (Area 4.1). Trace fine sand and vegetal inclusions, hip width 1.37, thickness 1.17, very pale brown (10YR 8/4) body (cf. Farukhābād B6 [Earlier Middle Elamite]: Pl. 20*E*; cf. Gautier and Lampre 1905: Figs. 122, 126 [Musiyān]).

i, Abstract figurine (Area 5.2). Trace fine sand inclusions, height 6.07, top width 5.30, top thickness 2.13, oval base 3.39 × 2.05, pale yellow (2.5Y 8/3) body.

j, Ring scraper (Area 5.2). 10% fine sand inclusions, Dm 10, height 3.1, RT 0.71, pink (7.5YR 8/5) body (Farukhābād B36 [Early Uruk]: Fig. 76*n*; Suse Ville Royale BVI [Šimaški]: Pl. 33:13; Suse Ville Royale I, Level 5 [VA]: Fig. 38:9).

k, Incised wheel (Area 3.2). 5% fine sand inclusions, Dm ca. 8.35, hub thickness 3.35, hole Dm 0.57, very pale brown (10YR 7/4) body.

The Archaeological Sites and Their Interpretation

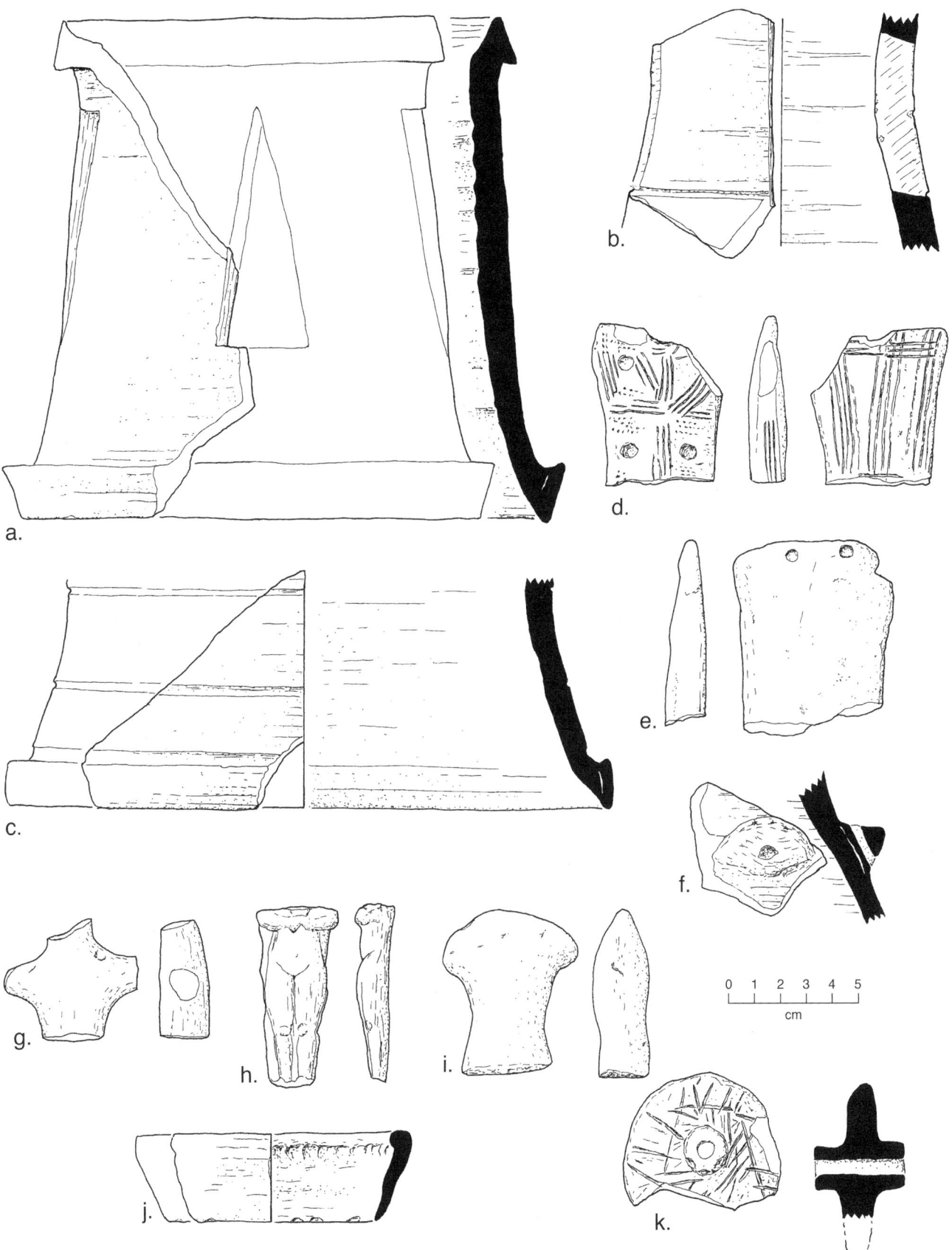

Figure 4.7. Varia from Musiyān (DL-20).

Figure 4.8. Late IIIrd and IInd millennium B.C. goblets, bases, and bowls from Musiyān (DL-20).

a, Goblet rim (SI-RG-1) (Area 3.1). No visible inclusions, Dm 6.5, NT 0.36, ST 0.45, greenish white (5Y 8/2) body, (Farukhābād B18 [Šimaški]: Fig. 90g; Suse Ville Royale BVI–V [Šimaški] Group 21a: Pl. 25:3, 12, 24, 26).

b, Goblet rim (SM-RG-1) (Area 4.3). 5% vegetal inclusions, Dm 8, NT 0.35, ST 0.39, very pale brown (10YR 8/4) body, pink (2.5YR 7/3) sludge (Farukhābād B13 [Sukkalmaḫḫu]: Fig. 85g; Suse Ville Royale Royale AXIV–XV [Sukkalmaḫḫu] Group 21ab: Pl. 23:6, 11, 12, 14, 15, 28, 31, 32, Pl. 24:4, 11, 17, 19, 21).

c, Goblet shoulder (Area 5.1). Trace vegetal and calcite inclusions, body Dm 9, NT 0.43, ST 0.94, reddish yellow (5YR 7/6) body (same comparanda as a).

d, Button base of goblet with ridge (SM-BG-2) (Area 5.1). Trace vegetal and calcite inclusions, base Dm, 4.5, ST 0.85, very pale brown (10YR 8/3) body (note pre-firing perforation) (Farukhābād B1–3 [Earlier Middle Elamite]: Fig. 89u).

e, Knob base of goblet (Area 5.1). Trace vegetal inclusions, base Dm 4.1, ST 0.50, light reddish brown (5YR 6/5) body, white (10YR 8/2) surface (Farukhābād B13 [Sukkalmaḫḫu]: Fig. 87a; Suse Ville Royale AXIV [Sukkalmaḫḫu] Group 19b: Pl. 20:2–7, 14).

f, Button base of goblet (SM-BG-1) (Area 5.1). Trace vegetal inclusions, base Dm 5.0, ST 0.58, white (2.5Y 8/2) body (Farukhābād B12 [Sukkalmaḫḫu]: Fig. 87b, c, B4 [Earlier Middle Elamite]: Fig. 89s, B7 [Earlier Middle Elamite]: Fig. 89r).

g, Disc base of jar (SM-B-1) (Area 2.1). 10% vegetal and medium sand inclusions, base Dm 8.5, white (10YR 8/2) body, reddish-brown (2.5YR 5/4) sludge (Farukhābād B17 [Šimaški]: Fig. 84o, B4 [Earlier Middle Elamite]: Fig. 89s, B7 [Earlier Middle Elamite]: Fig. 89r).

h, Disc base of jar (SI-B-1) (Area 5.1). 10% medium sand and calcite inclusions, base Dm 6.0, ST 1.01, light gray (10YR 7/1) body, interior damaged (Farukhābād B19 [Šimaški]: Fig. 84n, cf. B11–13 [Sukkalmaḫḫu]: Fig. 87j–L, B7 [Earlier Middle Elamite]: Fig. 89x).

i, Disc base of jar (SM-B-1) (Area 5.1). 5% vegetal inclusions, base Dm 7.0, ST 0.78, light reddish brown (2.5YR 6/4) body, very pale brown (10YR 8/3) surface, exterior heavily tooled (cf. Farukhābād B17 [Šimaški]: Fig. 84o, B12 [Sukkalmaḫḫu]: Fig. 87m).

j, Ring base (SI-B-2) (Area 5.1). 5% fine sand inclusions, base Dm 15, ST 1.15, pale yellow (2.5Y 8/3) body (Farukhābād B15–16 [Šimaški]: Fig. 87u; Suse Ville Royale BV [Šimaški] Group 36: Pl. 47:3, 4, Group 36a: Pl. 45:1–7).

k, High ring base (SM-B-3) (Area 3.2). 5% vegetal inclusions, base Dm 14, ST 0.88, greenish white (5Y 8/2) body (Farukhābād B13 [Sukkalmaḫḫu]: Fig. 87v, B1–3 [Earlier Middle Elamite]: Fig. 89y; Suse Ville Royale AXIV–XV [Sukkalmaḫḫu] Group 34: Pl. 43:1, 2, 6, 7, Group 36: Pl. 46:1–4).

l, Carinated bowl with band rim (Area 3.1). 5% medium sand and vegetal inclusions, Dm 24, RT 1.09, ST 0.89, very pale brown (10YR 8/3) body.

m, Incurved bowl with round base (Area 5.1). 10% fine sand, calcite and vegetal inclusions, rim Dm 9.5, body Dm 10.5, height 4.8, RT 0.21, ST 0.56, light red (2.5YR 6/7) body, red (2.5YR 5/7) surface.

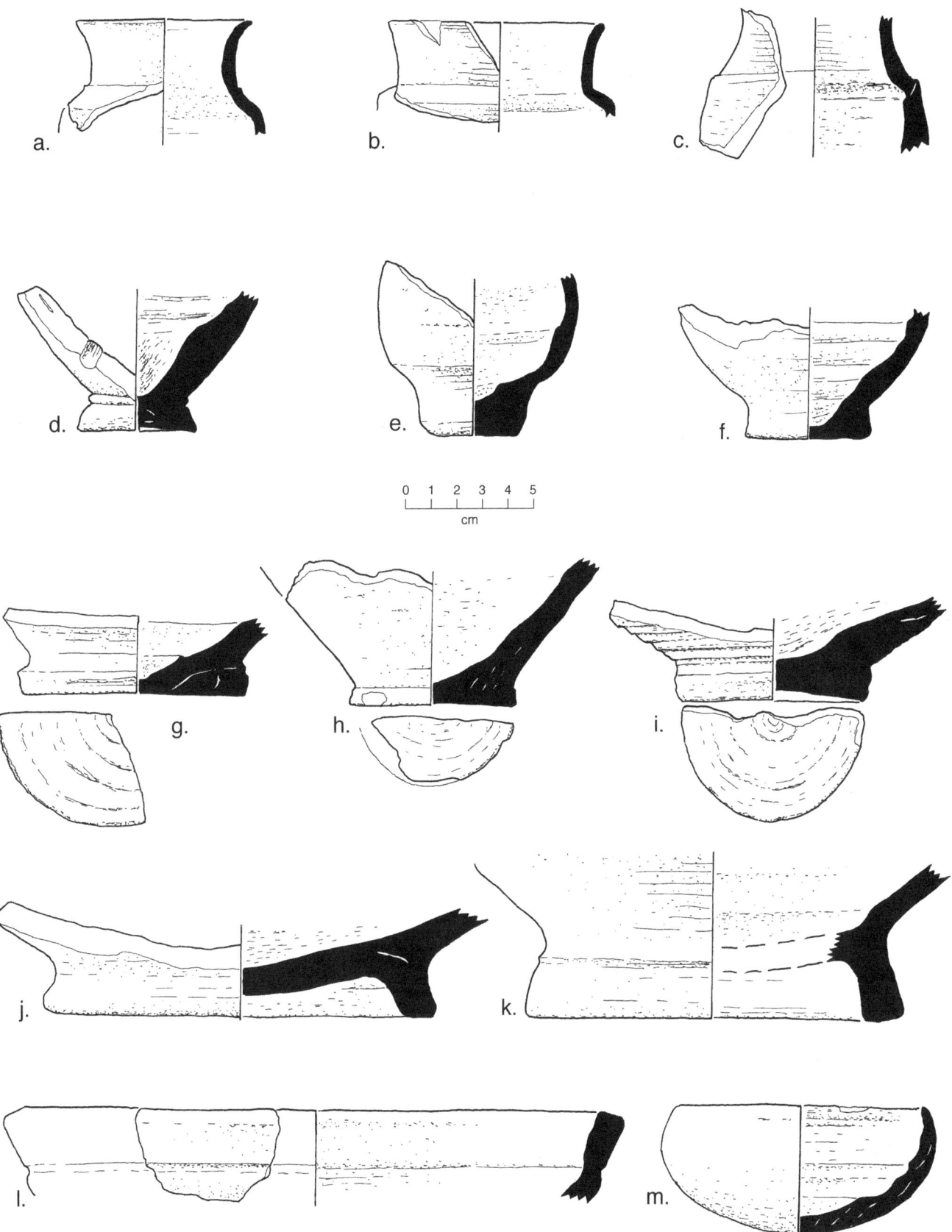

Figure 4.8. Late IIIrd and IInd millennium B.C. goblets, bases, and bowls from Musiyān (DL-20).

Figure 4.9. Late IIIrd and IInd millennium B.C. jars from Musiyān (DL-20).

a, Flared neck jar with thickened round rim (ED-RJ-1) (Area 5.1). 10% medium sand inclusions, Dm 14, RT 1.01, NT 0.75, white (2.5Y 8/2) body (post-firing scratches inside neck) (Farukhābād A3–5 [EDII+]: Fig. 49*b*, *L*).

b, Carinated jar with band rim (ED-RJ-6) (Area 3.2). Trace fine sand, calcite, and vegetal inclusions, Dm 12, body Dm 14.5, RT 0.45, NT 0.37, pink (7.5YR 7/4) body, faint traces of red painted rim band and shoulder triangles (Suse Ville Royale I, Level 7 [IVB]: Fig. 32:2, Level 6 [VA]: Fig. 39:1, Level 4b [VB]: Fig. 50:4, Level 3 [VB]: Fig. 46:12).

c, Jar with heavy band rim (SI-RJ-5) (Area 3.2). 5% medium sand inclusions, Dm 15, RT 1.07, NT 0.88, light red (2.5YR 6/6) body, white (10YR 8/2) surface (Suse Ville Royale BVI [Šimaški] Group 29b: Pl. 36:5, 8; Nippur WFIX–VIII [Type C13a: Akkadian-Ur III]: McMahon 2006:112, Pl. 105:2).

d, Jar with heavy ledge rim (SI-RJ-6) (Area 3.2). 10% medium sand and vegetal inclusions, Dm 18, RT 1.82, NT 1.02, RHt 1.39, color not recorded (cf. Suse Ville Royale BVI [Šimaški] Group 29b: Pl. 36:6; Suse Ville Royale I, Level 4 [VB]: Fig. 50:11; Nippur WFIX–VIII [Type C25: Ur III]: McMahon 2006:116, Pl. 119).

e, Jar with ledge rim (ED-RJ-4) (Area 5.1). 10% vegetal and medium sand inclusions, Dm 18, RT 1.55, NT 0.61, RHt 1.16, white (2.5Y 8/2) body (Farukhābād A13 [Jemdet Nasr]: Fig. 53*f*, *g*, B22 [EDII+]: Fig. 53*a*).

f, Jar with ledge rim (SI-RJ-2) (Area 5.1). 15% coarse sand and vegetal inclusions, Dm 18, RT 1.68, NT 0.93, RHt 1.105, white (2.5Y 8/2) body (pre-firing scratches inside neck) (Farukhābād B22 [EDII+]: Fig. 53*q*).

g, Jar with ledge rim (SM-RJ-1) (Area 5.1). 10% vegetal inclusions, Dm 27, RT 2.15, NT 0.75, RHt 1.53, pale yellow (2.5Y 8/3) body (Suse Ville Royale I, Level 7 [IVB]: Fig. 32:1).

h, Jar or stand with oblique rim (SM-RJ-3) (Area 5.1). 10% vegetal and fine sand inclusions, Dm 21, RT 1.65, NT 0.85, RHt 1.74, white (2.5Y 8/2) body (Farukhābād B15–17 [Šimaški]: Fig. 84*m*; Suse Ville Royale AXIII–B [Šimaški-Sukkalmaḫḫu]: Pl. 33:9–12).

i, Heavy ribbed bowl rim (Area 3.2). 5% vegetal and medium sand inclusions, Dm 20, RT 1.55, ST 1.20, RHt 1.47, white (2.5Y 8/2) body (Tell Asmar Houses VA [Protoimperial]: Pl. 105*e*, Delougaz 1952: Fig. 189, C.805.210).

j, Jar with oblique rim (SI-RJ-2) and grooved neck (Area 6.5w). Trace of fine sand, vegetal inclusions, Dm 18, RT 1.08, NT 0.60, light red (10R 6/5) body (cf. Farukhābād B13–16 [Šimaški-Sukkalmaḫḫu]: Fig. 85*L*).

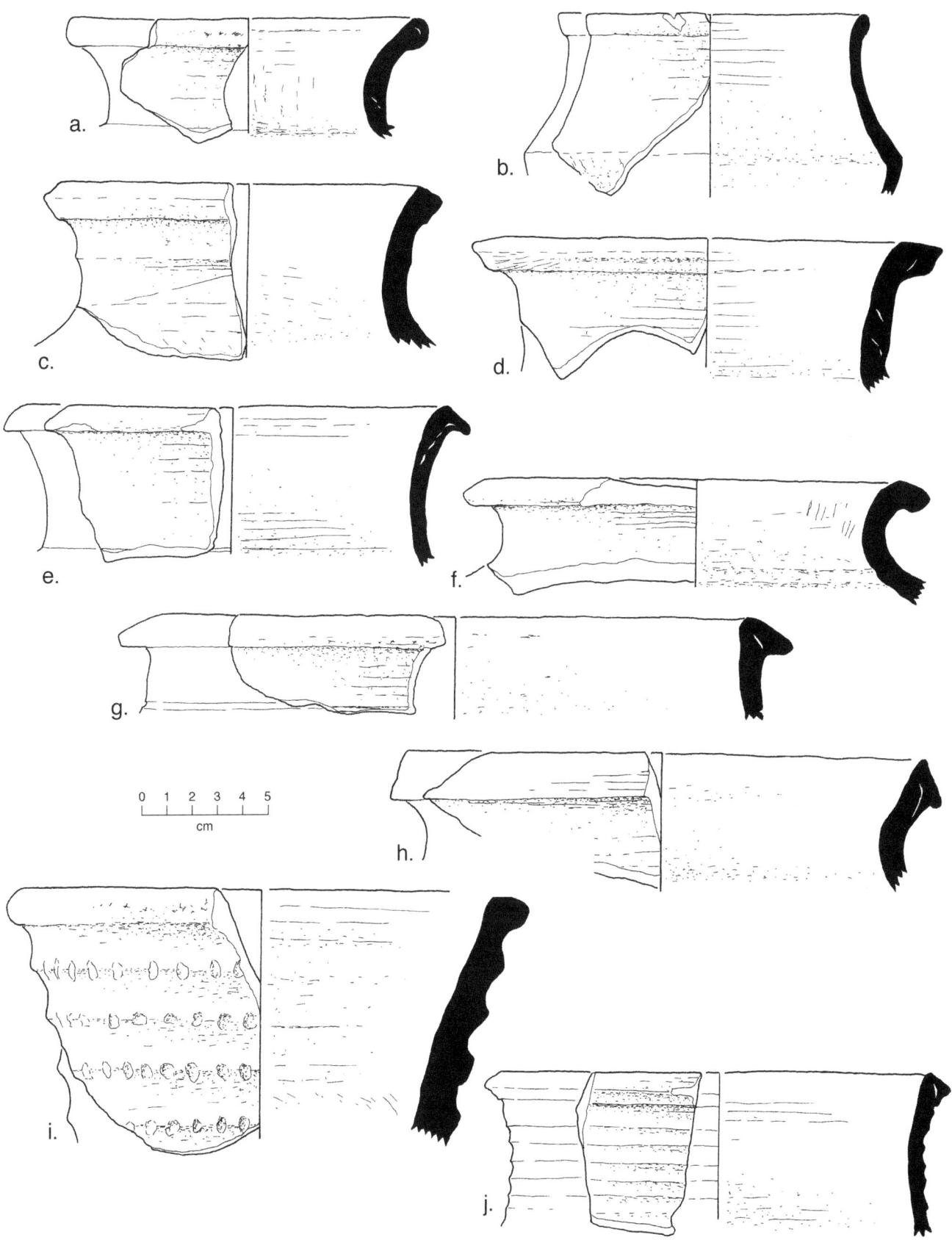

Figure 4.9. Late IIIrd and IInd millennium B.C. jars from Musiyān (DL-20).

Figure 4.10. Late IIIrd millennium B.C. jars and basins from Musiyān (DL-20).

a, Band rim jar (SI-RJ-3) with ribbed shoulder (SI-O-1) (Area 3.1). 5% fine sand and vegetal inclusions, Dm 16, RT 1.11, NT 0.61, ST 0.61, RHt 1.11, white (5YR 8/2) body (Suse Ville Royale I, Level 7 [IVB]: Fig. 32:1, Level 8 [IVB]: Fig. 35:3, Level 6 [VA]: Fig. 39:16).

b, Jar with indented band rim (SI-RJ-4) (Area 3.2). 15% fine sand and calcite inclusions, Dm 19, RT 1.42, NT 0.83, RHt 1.78, very pale brown (10YR 7/3) body, very pale brown (10YR 8/3) surface (Farukhābād B18 [Šimaški]: Fig. 84i; Suse Ville Royale BVII–VI [Šimaški] Group 16b: Pl. 16:12; Suse Ville Royale I, Level 3 [VB]: Fig. 50:12).

c, Jar with oblique rim (SI-RJ-2) (Area 1.1). 5% fine sand inclusions, Dm 11, RT 1.20, NT 0.78, RHt 1.62, white (2.5Y 8/2) body (Farukhābād A7 [EDI]: Fig. 51n, B16 [Šimaški]: Fig. 84L; Suse Ville Royale BVII–VI [Šimaški] Group 24b: Pl. 27:9–10).

d, Basin with grooved band rim (SI-RB-4) (Area 5.1). 5% fine sand and vegetal inclusions, Dm 23, RT 1.00, NT 0.78, RHt 2.65, pale yellow (2.5Y 8/3) body (Farukhābād B17–19 [Šimaški]: Fig. 84h).

e, Basin with grooved band rim (SI-RB-4) (Area 3.2). 10% fine sand inclusions, Dm 36, RT 1.24, NT 0.73, RHt 2.53, very pale brown (10YR 8/3) body (same comparanda as d).

f, Jar with grooved band rim (SI-RJ-6) (Area 5.1). 5% fine sand and calcite inclusions, Dm 15, RT 0.84, NT 0.82, RHt 2.69, very pale brown (10YR 7/3) body (Farukhābād B17–19 [Šimaški]: Fig. 84j; Suse Ville Royale BVI–V [Šimaški] Group 15b: Pl. 16:9–11; Suse Ville Royale I, Level 3 [VB]: Fig. 50:9, Level 4B [VB]: Fig. 50:5; cf. Nippur WFXI–VI [Ur III] McMahon 2006:116, Pl. 120:5–7).

g, Basin with grooved band rim (SI-RB-4) with wavy incising (SI-O-2) (Area 6.3c). 10% fine sand and calcite inclusions, Dm 34, RT 1.20, NT 1.01, RHt 3.40, white (2.5Y 8/2) body (same comparanda as d).

h, Basin with grooved band rim (SI-RB-4) with wavy incising (SI-O-2) (Area 3.2). 10% fine sand and calcite inclusions, Dm 34, RT 0.97, NT 0.90, RHt 4.30, white (2.5Y 8/2) body (same comparanda as d).

i, Basin with grooved band rim (SI-RB-4) (Area 6.1X). 10% fine sand and vegetal inclusions, Dm 33, RT 1.54, NT 0.57, RHt 3.70, white (2.5Y 8/2) body.

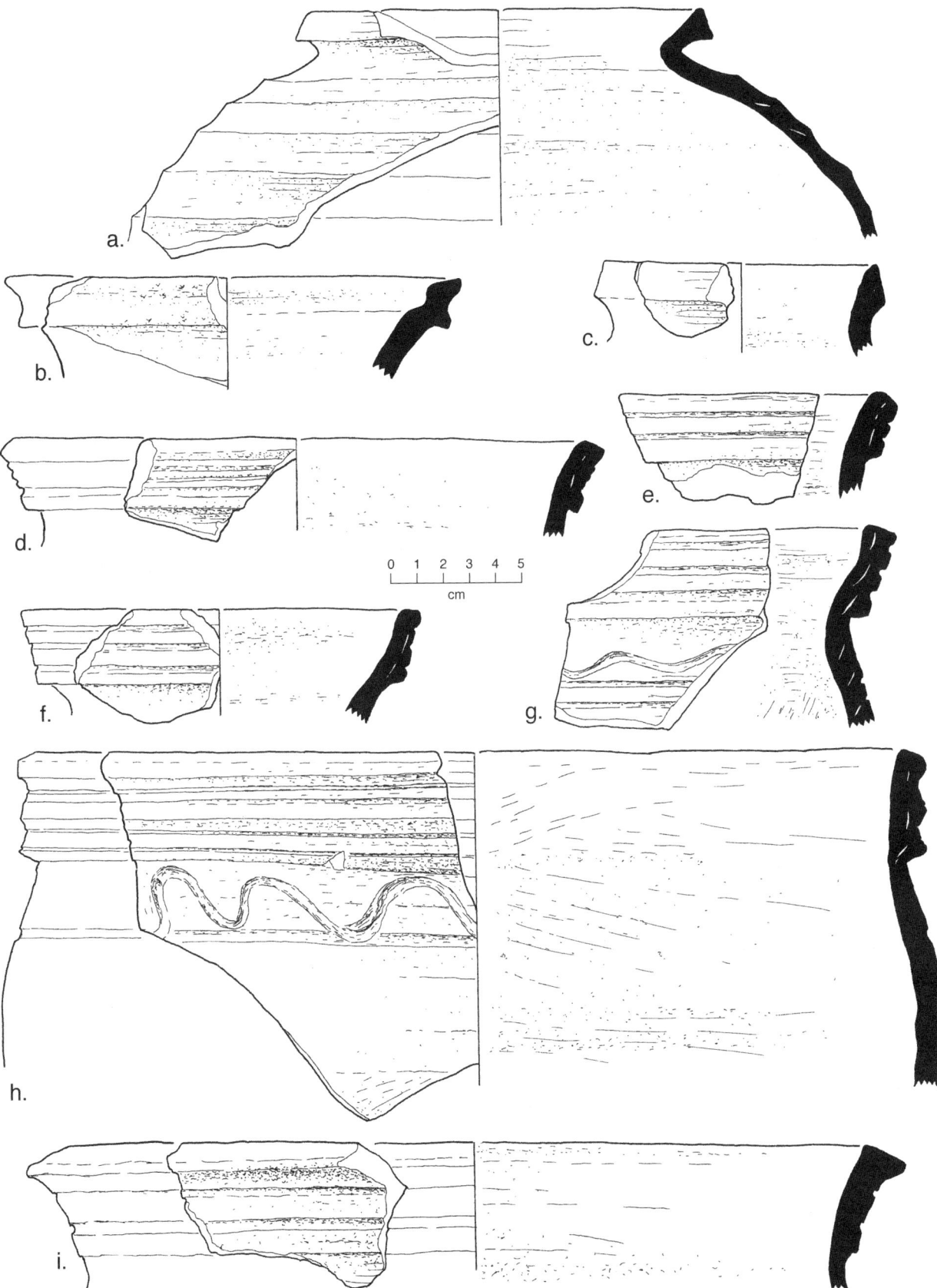

Figure 4.10. Late IIIrd millennium B.C. jars and basins from Musiyān (DL-20).

Figure 4.11. IInd millennium B.C. basin and bowl sherds from Musiyān (DL-20).

a, Large ribbed basin with flat lip (SI-RB-5) (Area 3.3). 10% medium sand and vegetal inclusions, Dm 38, RT 4.20, ST 1.35, white (2.5Y 8/2) body, interior damaged (Farukhābād B15 [Šimaški]: Fig. 90h, cf. B6 [Early Middle Elamite]: Fig. 89q; Suse Ville Royale I, Level 3 [VB]: Fig. 51:5).

b, Basin with ledge rim (SM-RB-5) (Area 5.1). 5% vegetal inclusions, Dm 38, RT 2.63, ST 1.24, RHt 2.06, light gray (2.5Y 7/2) body (Farukhābād B13–17 [Šimaški-Sukkalmaḫḫu]: Fig. 84g, B11–13 [Sukkalmaḫḫu]: Fig. 86h, j; Suse Ville Royale AXIV–XV [Sukkalmaḫḫu] Group 34: Pl. 43:1–9).

c, Basin with square rim (SM-RB-6) (Area 3.2). 10% vegetal inclusions, Dm ca. 58, RT 3.82, ST 1.25, RHt 2.60, color not recorded (Farukhābād B13 [Sukkalmaḫḫu]: Fig. 86n; Suse Ville Royale AXIV–XV [Sukkalmaḫḫu] Group 36a: Pl. 45:1–7; Suse Ville Royale I, Level 3–4A [VB]: Fig. 51:1–4).

d, Carinated bowl with square rim (Area 2.2). 15% medium sand inclusions, Dm 20, RT 1.30, ST 0.75, RHt 1.01, reddish-yellow (7.5YR 7/5) body, very pale brown (10YR 8/3) body (Suse Ville Royale I, Level 3, 4A [VB]: Fig. 49:14, 16) (the ware and rim treatment indicate this is Šimaški, but we find no close comparanda).

e, Ledge rim bowl (SI-RB-2) with wavy incising (SI-O-2) (Area 5.1). 5% medium sand inclusions, Dm 20, RT 1.74, ST 0.71, RHt 1.35, pink (7.5YR 7/4) body (Suse Ville Royale I, Level 4A [VB]: Fig. 49:18).

f, Jar shoulder with wavy incising (SI-O-2) (Area 5.1). Trace fine sand and calcite inclusions, ST 1.18, pinkish gray (7.5YR 7/3) body (Ville Royale I, Level 4A [VB]: Fig. 50:20, 21).

g, Square rim jar (SM-RJ-2) (Area 5.1). 5% vegetal inclusions, Dm 21, RT 1.91, NT 0.88, RHt 1.30, pale yellow (2.5Y 8/3) body (cf. Farukhābād B11–13 [Sukkalmaḫḫu]: Fig. 85n; Suse Ville Royale I, Level 4A [VB]: Fig. 49:18).

h, Ledge rim bowl with wavy incising (SI-RB-2) (Area 5.1). 5% fine sand inclusions, Dm 28, RT 2.22, NT 0.89, RHt 1.04, white (2.5Y 8/2) body (Suse Ville Royale I, Level 4A [VB]: Fig. 49:18).

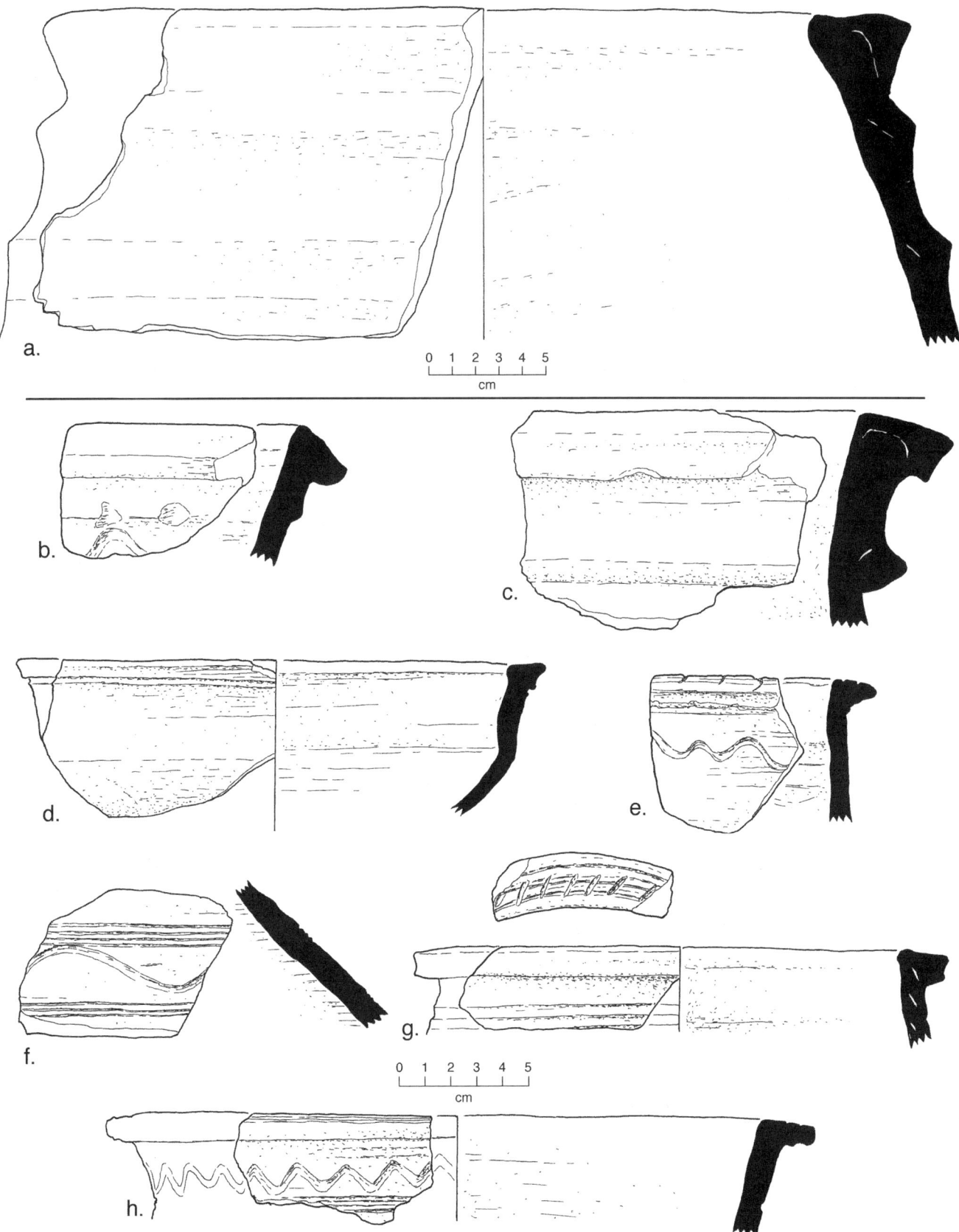

Figure 4.11. IInd millennium B.C. basin and bowl sherds from Musiyān (DL-20).

46 *Elamite and Achaemenid Settlement on the Deh Lurān Plain*

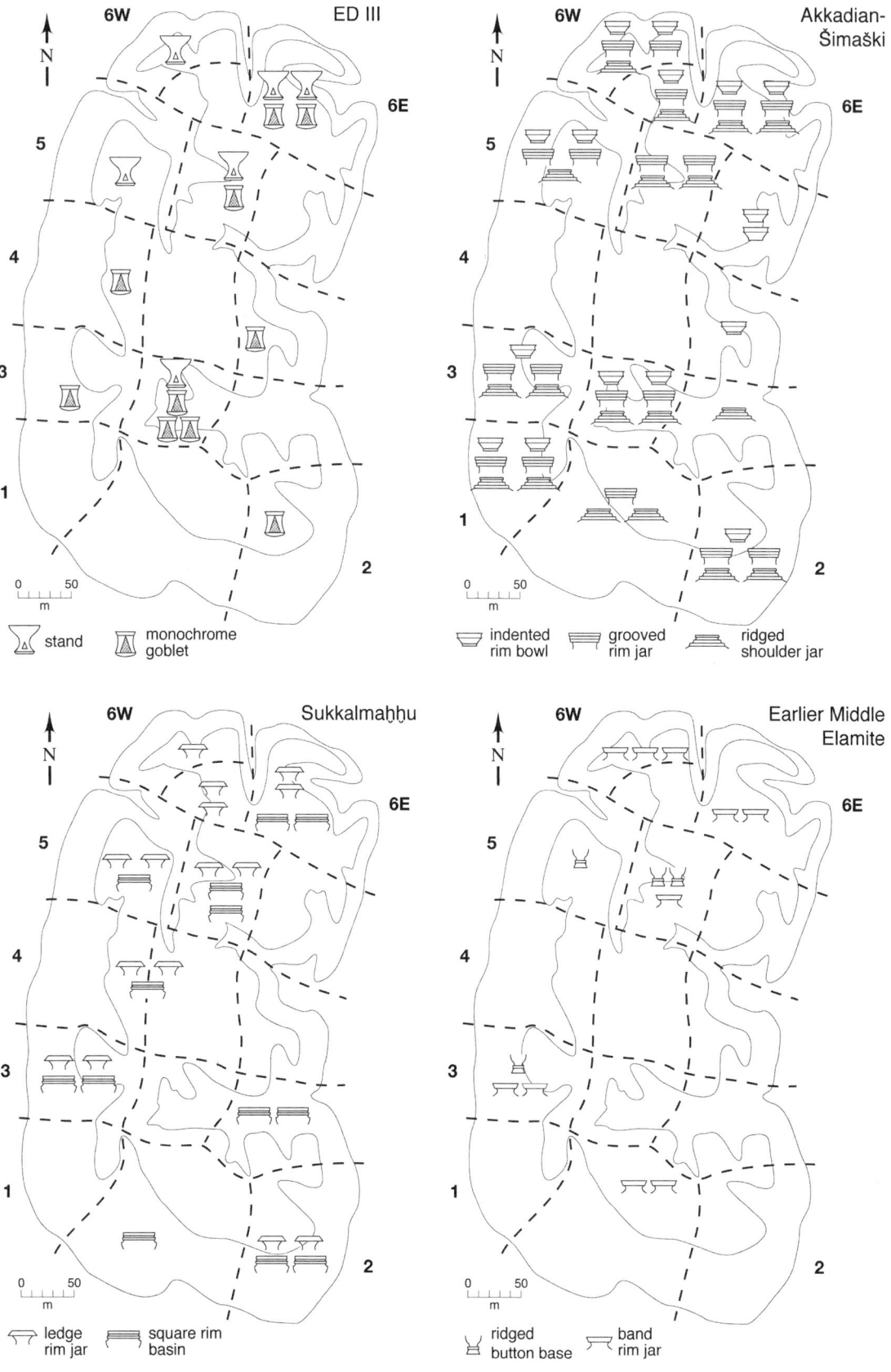

Figure 4.12. The distribution of IIIrd to Ist millennium B.C. ceramics on Musiyān (DL-20).

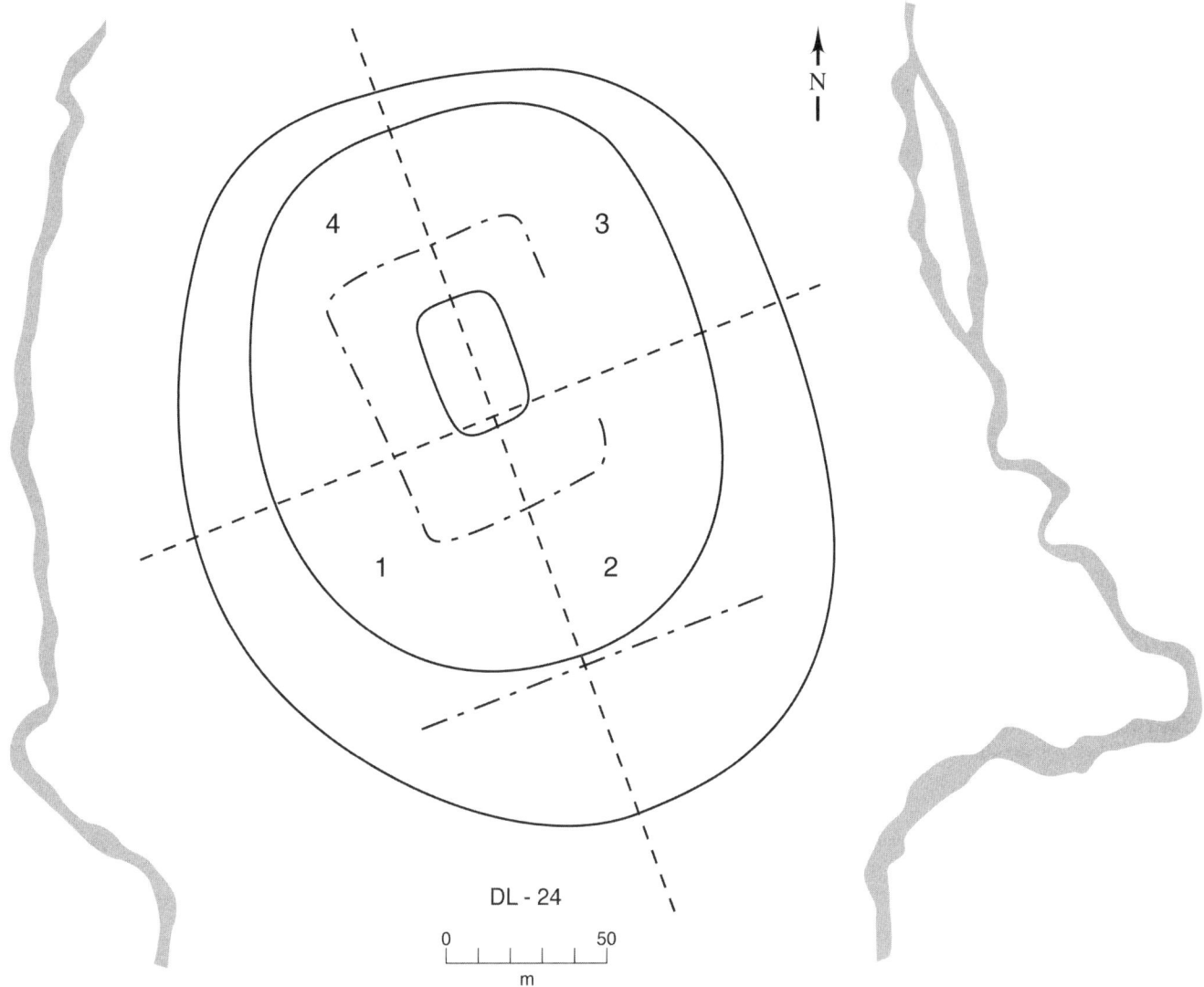

Figure 4.13. Map of Baulah (DL-24). (Gray lines mark recent watercourses; dot-dash lines mark possible mud brick traces.)

Number: DL-24
Name: Baulah
Grid: E 47°15'12" / N 32°39'55"
  NIOC: E 1704.6 – N 1182.6
  UTM: E 72, 704 / N 3, 604, 238
Dimensions:
  Length: 255 m (N-S)
  Width: 190 m (E-W)
  Height: 5.0 m
  Circumference: 725 m
  Area: 4.18 ha

Modern Environmental Context: On a terrace northeast of the Ab-i Garm stream where it leaves the northern alluvial slopes of the plain and spreads out into the saline depression, today an area of irrigated fields.

Features: The overall form of this site is elliptical (Fig. 4.13). Its outer limits are clearly defined by a small recent canal and the agricultural fields that surround it. This mound rises in four terraces. Either some special activity occurred on the site or some of the later construction phases of this site were burned, since gray ash and fire-cracked rocks are dense in the upper layers of the site and are being exposed in erosional cuts. Linear markings that may indicate walls, and are visible on recent satellite images, are shown on the map. A number of burials were found eroding from the northern portion of the second and third levels/terraces above the alluvial plain. Boulder wall footings were visible, scattered over the site's surface. A greenish-buff baked brick was found on the first terrace/level above the plain in the northwest quadrant of the site. It measured 20 by 22 cm and 5 cm in thickness, had rather irregular surfaces and edges, and showed evidence of much vegetal tempering. A rather large number of grinding stones and slabs, complete and fragmentary, were present on the surface. Many of these had apparently been exposed through the excavation of numerous shallow pits into the summit and upper levels/terraces of the site not long before our visit.

Phases: A moderate quantity of ceramics was found, with the greatest amount appearing in the southwest quadrant of the site and the least in the northeast. The major occupations are those of the IIIrd and IInd millennium. Early Dynastic I–II sherds were found over the entire site, with the greatest numbers and density occurring on the northern half of the site, as discussed in the earlier publication (Neely and Wright 1994:78). Rims (ED-RJ-6), monochrome decorated shoulders (ED-O-6: Neely and Wright 1994: Fig. IV.21o), and bases (ED-B-5) of the small carinated pots of the Early Dynastic III phase are also widely scattered.

The Elamite ceramics include several Šimaški phase diagnostics including a band rim jar (SI-RJ-3) with ribbed shoulder (SI-O-1: Fig. 4.14a) from the north half of the site and a distinctive jar with a thickened band rim (SI-RJ-5: Fig. 4.14b) from the south half of the site. Several other items with Šimaški parallels are similarly scattered (Fig. 4.14d, h). Sukkalmaḫḫu or earlier Middle Elamite ceramics include a band rim jar (SM-RJ-4: Fig. 4.14c), a ledge rim bowl (SM-RB-3: Fig. 4.14f), as well as a heavy ring base (Fig. 4.14i). These, too, are scattered on both the north and south portions of the site.

Comment: Though the few Elamite traces are obviously remnants of badly eroded layers or features, they are widely scattered and we have therefore assigned the full site area of 4.18 ha to the Šimaški and Sukkalmaḫḫu occupations. See Chapter 6 for a discussion of associated channel DL-276B.

Reference: Carter 1971:230.

Figure 4.14. Late IIIrd and IInd millennium B.C. jars, basins, and bases from Tepe Baulah (DL-24).

a, Jar with band rim (SI-RJ-3) and ribbed shoulder (SI-O-1) (Area 4). 15% medium sand inclusions, Dm 20, RT 1.49, NT 0.68, ST 0.83, RHt 1.35, pale yellow (2.5Y 8/3) body (Suse Ville Royale I, Level 7 [IVB]: Fig. 32:1, Level 8 [IVB]: Fig. 35:3).

b, Jar with thickened band rim (SI-RJ-5) (Area 2). 15% fine sand and calcite inclusions, Dm 15, RT 1.07, NT 0.69, RHt 1.10, pale yellow (5Y 7/3) body (Suse Ville Royale BVI [Šimaški] Group 29b: Pl. 36:5, 8).

c, Band rim jar (SM-RJ-4) (Area 4). 10% coarse vegetal and fine sand inclusions, Dm 18, RT 1.25, NT 0.75, RHt 2.20, pale yellow (5Y 8/3) body (Farukhābād B11–13 [Sukkalmaḫḫu]: Fig. 86b; Suse Ville Royale AXII–XIII [Sukkalmaḫḫu] Group 29a: Pl. 34:5, Pl. 35:3–7, Group 30c: Pl. 38:1–4).

d, Basin with square rim (SM-RB-6) (Area 2). 10% medium sand and vegetal inclusions, Dm 25, RT 2.67, NT 1.20, RHt 2.04, pale yellow (2.5Y 8/3) body (Farukhābād B15–16 [Šimaški]: Fig. 86e).

e, Flared neck jar with band rim (Area 3). 10% coarse vegetal and medium sand inclusions, Dm 12, RT 1.16, NT 0.64, RHt 1.87, greenish white (5Y 8/2) body (Farukhābād B11–13 [Sukkalmaḫḫu]: Fig. 85n).

f, Bowl with ledge rim (SM-RB-3) (Area 2). 5% vegetal inclusions, Dm ca. 29, RT 2.23, NT 0.91, RHt 1.60, light red (2.5YR 6/6) body (Farukhābād B11–13 [Sukkalmaḫḫu]: Fig. 85m).

g, Ring base with impressions (ED-B-7). 15% medium sand inclusions, base Dm 16, ST 0.90, dark brown (10YR 6/3) body (Suse Ville Royale I, Level 12 [IVA]: Fig. 27:3, Level 9B [IVA]: Fig. 27:5, 7).

h, Ceramic wheel (Area 3). 20% medium sand inclusions, Dm ca. 8.5, hub thickness 6.25, pale yellow (2.5Y 7/4) body, very pale brown (10YR 8/3) surface (Suse Ville Royale I, Level 8 [IVB]: Fig. 36:2).

i, Heavy ring base (Area 2). 5% vegetal inclusions, base Dm 22, ST ca. 1.65, white (2.5Y 8/2) body (the ware indicates this is Sukkalmaḫḫu or earlier Middle Elamite, but we find no exact comparanda).

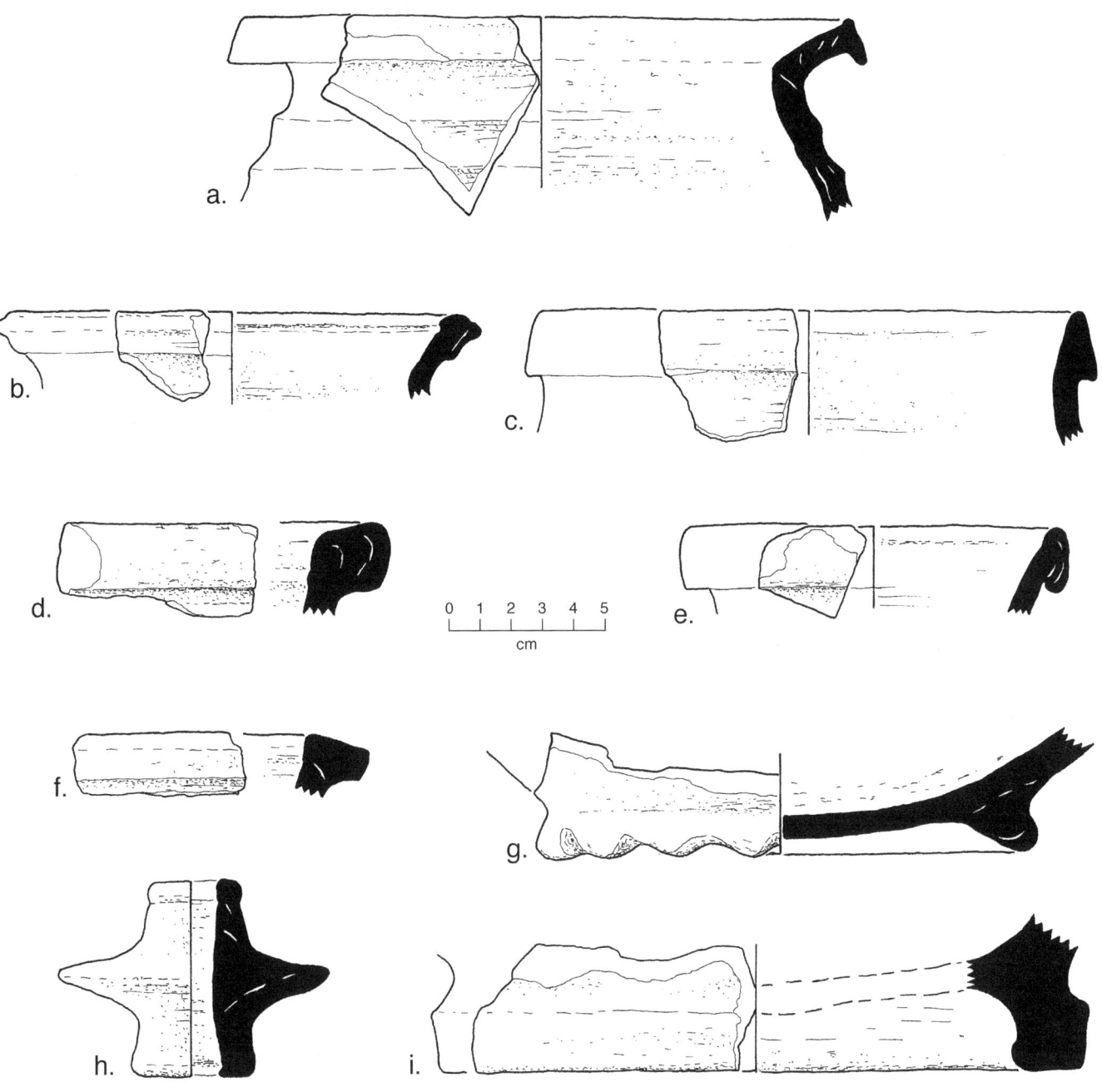

Figure 4.14. Late IIIrd and IInd millennium B.C. jars, basins, and bases from Tepe Baulah (DL-24).

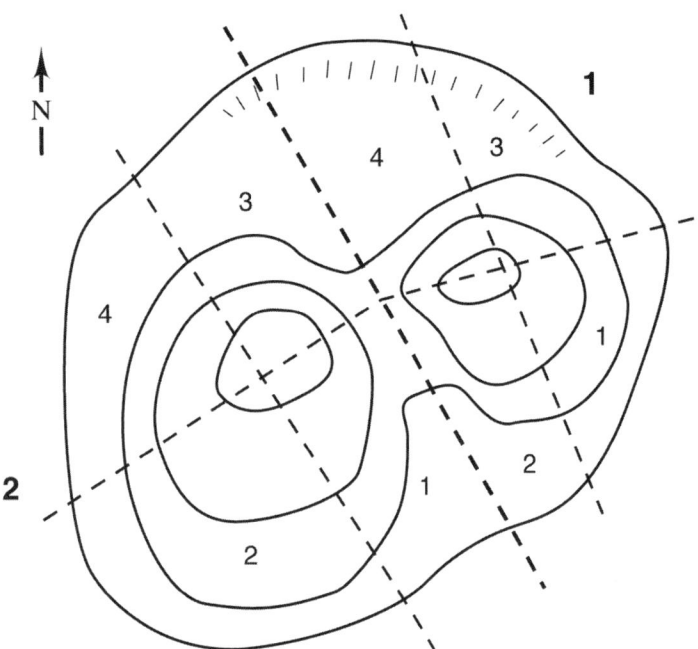

Figure 4.15. Map of Tenel Ramon (DL-27).

Number: DL-27
Name: Tenel Ramon
Grid: E 47°11'04" / N 32°38'00"
 NIOC: E 1704.6 – N 1182.8
 UTM: E 705,011 / N 3,612,225
Dimensions:
 Length: 170 m (E-W)
 Width: 150 m (N-S)
 Height: West: 8.0 m; East 7.0 m
 Circumference: ca. 600 m
 Area: 2.49 ha

Modern Environmental Context: On the Mehmeh River alluvial fan, about 300 m southwest of the present channel of the river.

Features: The site consists of a large oval mound having two distinct parts or "lobes" separated by a lower "saddle" (Fig. 4.15). The larger western lobe of the mound measures about 90 m east-west by 110 m north-south at the present level of the alluvial plain. The sides of the western lobe rise in an unbroken steep slope to a relatively broad, slightly mounded, summit about 8 m above the present level of the alluvial fan. This summit measures 38 m east-west by 21 m north-south. The smaller eastern lobe measures some 72 m east-west and 60 m north-south at its base. The slightly convex summit rises about 7 m above the plain, and measures 25 m east-west by 12 m north-south. A terraced area, about 5.5 m above the plain, extends from the steeply sloping sides of the summit toward the east. The summits of both lobes have been rather badly disturbed by modern burials and what appear to be looter's holes.

Two distinct, possibly architectural, terraces extend in a steplike fashion from the north side of the bi-lobed mound. The uppermost of the two stepped terraces to the north is about 2.5 m above the plain at its juncture with the bi-lobed mound. The lower terrace is about 1.5 m above the plain at its juncture with the upper terrace. The northern end of the lower terrace blends into the surrounding alluvium.

Erosional processes on the summits and all slopes of the site are exposing a large number of cobble and boulder wall footings. Several of the footings on the summits may be recent, but others are undoubtedly contemporaneous with the site's ancient occupations. Erosion has also exposed small portions of ceramic coffins on the southwest slope of the western lobe, at levels between 3 and 4 m below its summit. Unfortunately, not enough of these coffins were exposed to permit an assessment of their age.

Recent satellite images show that land leveling for agriculture around the site has trimmed all sides, particularly the low, seemingly terraced north side.

Phases: A moderate to heavy sherd concentration was scattered over the surface of the site. Early Dynastic I–II occupation is indicated by a solid footed conical cup base (ED-CB-1), thin round lip jars (ED-RJ-2), and jar shoulders with polychrome motifs (ED-O-5) scattered on and around the mound (Neely and Wright 1994: Fig. 84).

Many Elamite sherds were found in all areas of Tenel Ramon (Figs. 4.16–4.18). There is one item that is definitely Šimaški, a basin with grooved rim (SI-RB-4: Fig. 4.18b). Still others are probably of Šimaški date (Figs. 4.16m, 4.17e, 4.18c, e). Probable Šimaški items are predominantly in Area 1. Almost all of the rest of the sherds were from vessel types characteristic of the Sukkalmaḫḫu and Earlier Middle Elamite periods, and it is likely that thick deposits of these periods mask earlier material. Many forms were made in both of the later Elamite periods. Sherds of goblet rims and shoulders (SM-RG-1: Fig. 4.16*a–c*, *e*), flat bases (SM-BG-2: Fig. 4.16*h*), button bases (SM-BG-4: Fig. 4.16*d*, *f*, *g*), and knob bases (Fig. 4.16*j*) are widespread. Various simple small bowls (SM-RB-1: Fig. 4.16*n*, *o*) are found. Similarly, various ledge rim jars (SM-RJ-1: Fig. 4.17*c*), band rim jars (SM-RJ-4: Fig. 4.17*h*, *i*), and square rim jars (SM-RJ-2: Fig. 4.17*l*) are common. Basins with flat rims (SM-RB-4: Fig. 4.18*e*) continue into these periods, as do basins with square rims (SM-RB-6: Fig. 4.18*a*, *g*). Vessels characteristic of Sukkalmaḫḫu including bowls with outturned rims (SM-RB-2: Fig. 4.16*p*), jars with oblique ledge rims (SM-RJ-1: Fig. 4.17*f*), and basins with band rims (SM-RB-7: Fig. 4.18*d*) were found on all parts of the site. Vessels characteristic of Earlier Middle Elamite include the flat goblet bases with ridge (SM-BG-2: Fig. 4.16*h*) and indented band rim jars (SM-RJ-5: Fig. 4.17*j*, *k*). These too are found scattered on all parts of the mound.

The few later items will be reported in a subsequent study.

Comment: There is no definite indication of occupation in Later Early Dynastic III or Awan times. Šimaški remains appear to have a limited distribution and we ascribe 1.0 ha to this occupation. Sukkalmaḫḫu Elamite and Earlier Middle Elamite diagnostics indicate the entire mound was occupied, and we ascribe an area of 2.0 ha to these occupations. See Chapter 6 for a discussion of associated Canals DL-319 and 320.

Reference: Carter 1971:230–31.

Number: DL-32
Name: Tepe Farukhābād
Grid: E 47°13'08" / N 32°35'18"
  NIOC: E 1708.5 – N 1177.9
  UTM: E 708, 259 / N 3, 607, 823
Dimensions:
  Length: Entire site is now 207 m (NW-SE); Central Mound now 145 m (NW-SE), formerly about 200 m.
  Width: Entire site is now 140 m (NE-SW); Central Mound now 90 m (NE-SW), formerly about 200 m.
  Height: 25 m above level of the old alluvium.
  Circumference: Entire site now 620 m; Central Mound now 395 m, formerly about 520 m.
  Area: Entire site now 2.40 ha; Central Mound now 0.96 ha, formerly about 3.0 ha.

Modern Environmental Context: On the Mehmeh River alluvial fan, directly adjacent to the present channel of the river.

Features: DL-32 is one of most impressive sites on the Deh Lurān Plain. Today, the highest part of Farukhābād appears as a steep-sided mound forming the western part of the site (Fig. 4.19, Plate 3). About half way down the eastern and southeastern slopes, a broad apron-like extension of the site drops northeastward to blend into the alluvial plain (Wright 1981: Fig. 3, Plate 2*A*). The very steep southwestern face of the mound (Wright 1981: Plate 2*B*; Hole, Flannery, and Neely 1969: ii, frontispiece) is a result of long-term erosion and undercutting by the Mehmeh River. Wright (1981:4) estimates that the large, steep mound was originally about 200 m in diameter, and that perhaps as much as 60 percent of it has been eroded away. Wright also observes that the erosion of the Mehmeh River has been downward as well as lateral in nature. He notes that while the present summit of the great mound rises about 30 m above the current floodplain of the Mehmeh, the ancient floodplain of the river was approximately 5 m higher. Wright's 1968 excavations did not attempt to reach the sterile ground surface on which the site was founded. However, the face cut by the river showed the original ground surface to be about 5 m above the ancient flood plain. Thus, the maximum cultural accumulation was about 25 m.

Phases: Wright's (1981) report on the 1968 excavations at Tepe Farukhābād provides detailed information pertaining to the architectural features and artifacts disclosed, which need not be recapitulated here. The fieldwork and analysis revealed evidence that the site had been discontinuously occupied during fifteen cultural phases. By the Early Dynastic III phase (Wright 1981:164–94), the mound reached an elevation of 156 m above sea level or 22 m above the plain level at the time of occupation. It had probably diminished somewhat from its Early Dynastic I–II size, but still had both elaborate and modest housing. Farukhābād was evidently then abandoned until the Šimaški Elamite phase, about 2100 B.C., when it was partially occupied by a village of modest housing. Elamite occupation of the site continued through most of the following Sukkalmaḫḫu phase (Carter 1981:200–223), but it was much smaller and centered, at least late in this period, on a small, probably oval, fortification. At about 1600 B.C. this was followed by a similarly small earlier Middle Elamite phase occupation composed of modest housing. This phase ended at about 1500 B.C., bringing the mound to a little more than 160 m above sea level, 26 m above plain level.

The latest occupation at Farukhābād disclosed by Wright's excavations dated to the Late Parthian and Early Sasanian periods (ca. 50 B.C. to A.D. 350). This last use of Farukhābād, centered on the lower, broad, apron-like terrace forming the eastern and southeastern portions of the site (Carter 1981:224–26), will be discussed in our final survey volume.

Comment: We estimate that Jemdet Nasr and Early Dynastic I–II occupation covered an area of about 3.0 ha, and that the Early Dynastic III occupation was somewhat smaller, perhaps 2.5 ha, since the mound's summit had diminished in area as it grew higher. The Šimaški phase occupation seems to have been a small village covering only part of the summit of the mound, perhaps 0.50 ha. The Sukkalmaḫḫu settlement had a small fortified installation covering about 0.2 ha. This developed into a small hamlet of similar size in earlier Middle Elamite times.

References: Gautier and Lampre 1905:84; Carter 1981:196–226; Wright 1981.

Figure 4.16. IInd millennium B.C. jars and bases from Tene1 Ramon (DL-27).

a, Goblet rim (SI-RG-1 or SM-RG-1) (Area 1.4). Trace fine sand inclusions, Dm 10, NT 0.43, very pale brown (10YR 8/4) body.

b, Goblet rim (same) (Area 1.4). Trace fine sand and vegetal inclusions, Dm 10, NT 0.47, pale yellow (2.5Y 8/3) body.

c, Goblet rim (SM-RG-1) (Area 2.1). 5% vegetal inclusions, Dm 10, RT 0.28, NT 0.40, ST 0.49, very pale brown (10YR 8/3) body (Farukhābād B12–13 [Sukkalmaḫḫu]: Fig. 85f, g; Suse Ville Royale AXIV–XV [Sukkalmaḫḫu] Group 19b: Pl. 20:7–15, 17–20).

d, Button base of goblet (SM-BG-4) (Area 2.1). 5% vegetal and fine sand inclusions, base Dm 4.18, ST 0.59, white (2.5Y 8/2) body (Farukhābād B12 [Sukkalmaḫḫu]: Fig. 87m).

e, Goblet neck (Area 2.1). 5% vegetal inclusions, neck Dm 10, NT 0.32, ST 0.57, white (10YR 8/2) body (Farukhābād B15–16 [Šimaški]: Fig. 85i, B13 [Sukkalmaḫḫu]: Fig. 85h).

f, Button base of goblet (SM-BG-4) with inner plug (Area 2.1). 5% vegetal inclusions, base Dm 4.4, ST 0.65, pale yellow (2.5Y 8/3) body, base smooth, note chipping and asymmetry (Farukhābād B11–13 [Sukkalmaḫḫu]: Fig. 87j–L, B4–7 [Earlier Middle Elamite]: Fig. 89r, s).

g, Button base of goblet (SM-BG-4) (Area 1.4). 5% vegetal inclusions, base Dm 4.0, ST 0.71, pale yellow (2.5Y 8/3) body (Farukhābād B13–17 [Šimaški-Sukkalmaḫḫu]: Fig. 87f, B4–7 [Earlier Middle Elamite]: Fig. 89r, s).

h, Flat goblet base with ridge (SM-BG-2) (Area 1.4). 10% vegetal inclusions, base Dm 7.0, ST 0.84, light brown (7.5YR 6/4) body (Farukhābād B12 [Sukkalmaḫḫu]: Fig. 87m).

i, Flat goblet base with ridge (SM-BG-2) (Area 1.4). 15% coarse sand inclusions, base Dm 5.0, ST 0.98, pinkish gray (7.5YR 7/3) body, very pale brown (10YR 8/3) surface (Farukhābād B10 [Earlier Middle Elamite]: Fig. 89t).

j, Knob base of goblet (Area 2.1). 10% medium sand inclusions, base Dm 1.1, ST 0.43, white (10YR 8/2) body (Farukhābād B13 [Sukkalmaḫḫu]: Fig. 87a).

k, Flat base of jar (SI-B-1) (Area 1.4). 10% medium sand and vegetal inclusions, base Dm 8, ST 0.90, pale yellow (2.5Y 4/2) body (Farukhābād B17 [Šimaški]: Fig. 84o, cf. B12 [Sukkalmaḫḫu]: Fig. 87m).

l, Ring base of jar base (SM-B-3) (Area 1.4). 5% vegetal and fine sand inclusions, base Dm 8, ST 0.88, light brown (7.5YR 6/4) body.

m, Cylindrical vessel base (Area 1.4). Trace vegetal inclusions, base Dm ca. 7, ST 0.53, light reddish-brown (2.5YR 6/4) body (Suse Ville Royale I, Level 3 [VB]: Fig. 51:6).

n, Bowl with round lip (SM-RB-1) (Area 2.1). 5% vegetal and fine sand inclusions, Dm 15, RT 0.82, ST 0.58, pale yellow (2.5Y 8/3) body (cf. Farukhābād B18 [Šimaški]: Fig. 84d, B12 [Sukkalmaḫḫu]: Fig. 85c, B11u [Earlier Middle Elamite]: Fig. 89a; Suse Ville Royale II, Level 10 [Later Middle Elamite]: Fig. 11:7).

o, Bowl rim with thickened round lip (SM-RB-1) (Area 1.4). 10% vegetal inclusions, Dm 13, RT 0.76, ST 0.65, color not recorded (Suse Ville Royale BVII–VI [Šimaški] Group 5a: Pl. 6:6, 10).

p, Bowl with out-turned rim (SM-RB-2) (Area 1.4). 5% vegetal inclusions, Dm 21, ST 0.70, red (10R 6/6) body, pinkish gray (7.5YR 7/3) surface (Farukhābād B12 [Sukkalmaḫḫu]: Fig. 85b; Suse Ville Royale AXIII–XV [Sukkalmaḫḫu] Group 4bc: Pl. 4:7, 14–16).

q, Carinated ledge rim bowl (SM-RB-5) (Area 1.4). 10% vegetal and medium sand inclusions, Dm 34, RT 1.90, ST 0.65, RHt 1.01, greenish white (5Y 8/2) body (cf. Farukhābād B11–12 [Sukkalmaḫḫu]: Fig. 86g).

r, Basin rim (cf. SI-RB-5) (Area 1.4). 5% vegetal inclusions, Dm ca. 50, RT 2.43, ST 1.45, reddish-brown (2.5YR 6/4) body, very pale brown (10YR 7/3) surface.

s, Carinated bowl rim (cf. SM-RB-5) (Area 1.4). 5% vegetal and medium sand inclusions, Dm ca. 30, RT 1.90, ST 0.86, RHt 1.62, pinkish-gray (7.5YR 7/3) body, very pale brown (10YR 8/3) surface (Farukhābād B11–13 [Sukkalmaḫḫu]: Fig. 86h, m; cf. Suse Ville Royale BV [Šimaški] Group 35b: Pl. 44:7).

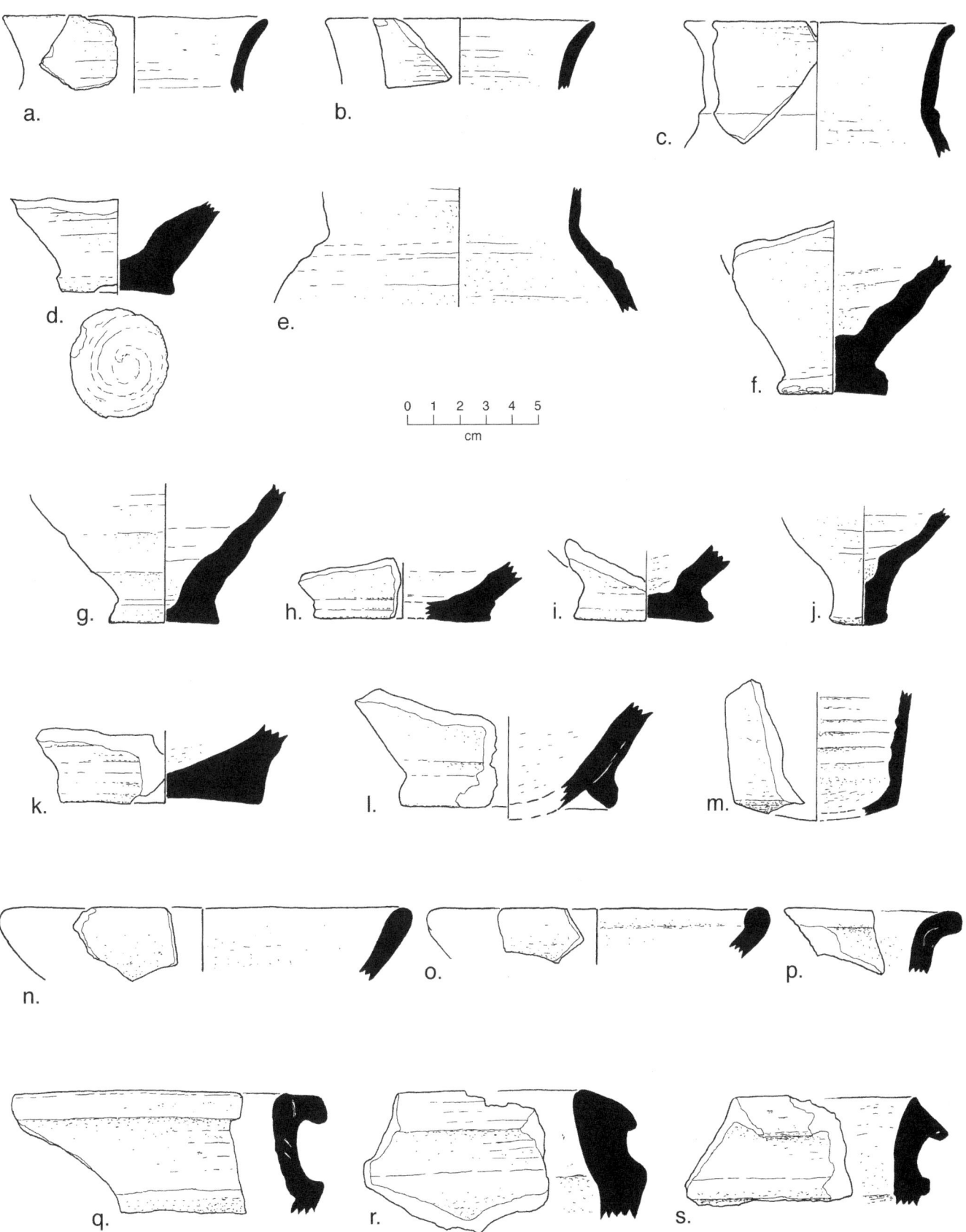

Figure 4.16. IInd millennium B.C. jars and bases from Tenel Ramon (DL-27).

Figure 4.17. Jar sherds and figurine from Tenel Ramon (DL-27).

a, Flared rim jar (Area 2.1). 5% fine sand and vegetal inclusions, Dm 16, NT 0.72, pinkish-gray (7.5YR 7/3) body, white (10YR 8/2) surface.

b, Flared goblet rim (SM-RG-1) (Area 2.1). 5% vegetal inclusions, Dm ca. 11, RT 1.14, NT 0.70, reddish-yellow (5YR 7/5) body (Farukhābād B12–13 [Sukkalmaḫḫu]: Fig. 85*e–g*, B8 [Earlier Middle Elamite]: Fig. 89*d*).

c, Ledge rim jar (SM-RJ-1) (Area 2.1). 10% vegetal inclusions, Dm 24, NT 0.85, reddish-brown (2.5YR 6/4) body, white (10YR 8/2) surface (Farukhābād B4–7 [Earlier Middle Elamite]: Fig. 89*L, n*; Suse Ville Royale AXII [Sukkalmaḫḫu] Group 30b: Pl. 37:4).

d, Jar with oblique rim (SM-RJ-3) (Area 1.1). 5% vegetal inclusions, Dm 18, RT 1.36, NT 1.03, very pale brown (10YR 8/3) body (Farukhābād B11–13 [Sukkalmaḫḫu]: Fig. 86*b*).

e, Band rim jar (SI-RJ-3) (Area 1.1). 5% medium sand and vegetal inclusions, Dm 17, RT 1.33, NT 0.70, white (2.5Y 8/2) body, very pale brown (10YR 8/4) surface (cf. Farukhābād B16–18 [Šimaški]: Fig. 84*i, L*; Suse Ville Royale I, Level 7 [IVB]: Fig. 32:1).

f, Ledge rim jar (SM-RJ-1) (Area 2.2). 5% vegetal and medium sand inclusions, Dm 16, RT 1.35, NT 0.50, white (2.5Y 8/2) body (Farukhābād B18 [Šimaški]: Fig. 84*k*, B11–13 [Sukkalmaḫḫu]: Fig. 86*m*; Suse Ville Royale AXIII [Sukkalmaḫḫu] Group 30a: Pl. 37:2).

g, Ledge rim jar (SM-RJ-1) (Area 2.1). 10% vegetal and medium sand inclusions, Dm 16, RT 1.76, NT 0.80, pale yellow (2.5Y 8/3) body (Farukhābād: B11–12 [Sukkalmaḫḫu]: Fig. 86*a*).

h, Band rim jar (SM-RJ-4) (Area 1.4). 10% vegetal and coarse sand inclusions, Dm 15, RT 1.18, NT 0.70, RHt 1.62, pale yellow (2.5Y 8/3) body (Farukhābād B15–17 [Šimaški]: Fig. 84*m*, cf. B11–13 [Sukkalmaḫḫu]: Fig. 86*b*).

i, Band rim jar (SM-RJ-4) (Area 1.4). 10% vegetal and fine sand inclusions, Dm 15, RT 1.26, NT 0.75, white (2.5Y 8/2) body (Farukhābād B15–17 [Šimaški]: Fig. 84*m*, B14 [Sukkalmaḫḫu]: Fig. 86*d*).

j, Indented band rim (SM-RJ-5) jar (Area 2.2). 5% medium sand and vegetal inclusions, Dm 13, RT 0.99, NT 0.70, pale yellow (2.5Y 8/3) body (Farukhābād B1–3 [Earlier Middle Elamite]: Fig. 89*h, I*; Suse Ville Royale AIX–XI [Earlier Middle Elamite] Group 17a: Pl. 18:3, 4, AXII [Sukkalmaḫḫu] Group 29a: Pl. 35:1, Group 30a: Pl. 37:1).

k, Indented band rim (SM-RJ-5) jar (Area 1.4). Trace vegetal and fine sand inclusions, Dm 12, RT 1.12, NT 0.77, pink (7.5YR 7/4) body, white (10YR 8/2) surface (same comparanda as *j*).

l, Square rim jar (SM-RJ-2) (Area 1.4). 5% vegetal and fine sand inclusions, Dm 19, RT 1.89, NT 1.35, greenish white (5Y 8/2) body (Farukhābād B11–13 [Sukkalmaḫḫu]: Fig. 85*L*, B4 [Earlier Middle Elamite]: Fig. 89*k*).

m, Heavy flat rim jar (Area 1.1). 10% vegetal inclusions, Dm 15, RT 2.54, NT 1.19, colors not recorded.

n, Combed sherd (SI-O-3) (Area 1.1). 10% medium sand and vegetal inclusions, ST 1.13, greenish white (5Y 8/2) body (cf. Farukhābād B15–16 [Šimaški]: Fig. 90*f*; Suse Ville Royale AXIV [Sukkalmaḫḫu] Group 34: Pl. 43:1).

o, Combed sherd (SI-O-3) (Area 1.1). 5% vegetal and medium sand inclusions, ST 1.45, white (2.5Y 8/2) body (Suse Ville Royale AXV [Sukkalmaḫḫu] Group 34: Pl. 43:5).

p, Female figurine (Area 2.2). 5% medium sand, ST 1.17, very pale brown (10YR 8/3) body.

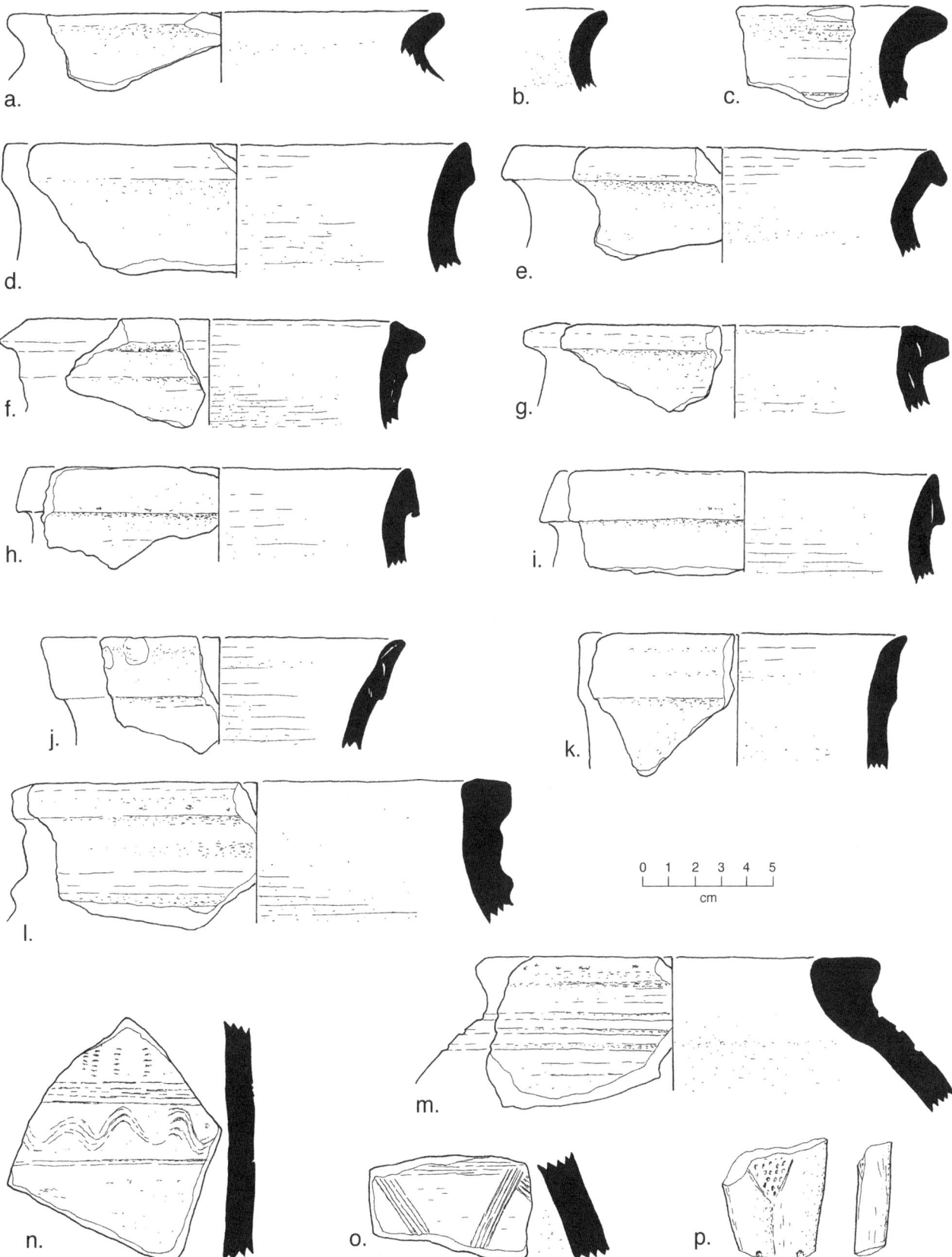

Figure 4.17. Jar sherds and figurine from Tenel Ramon (DL-27).

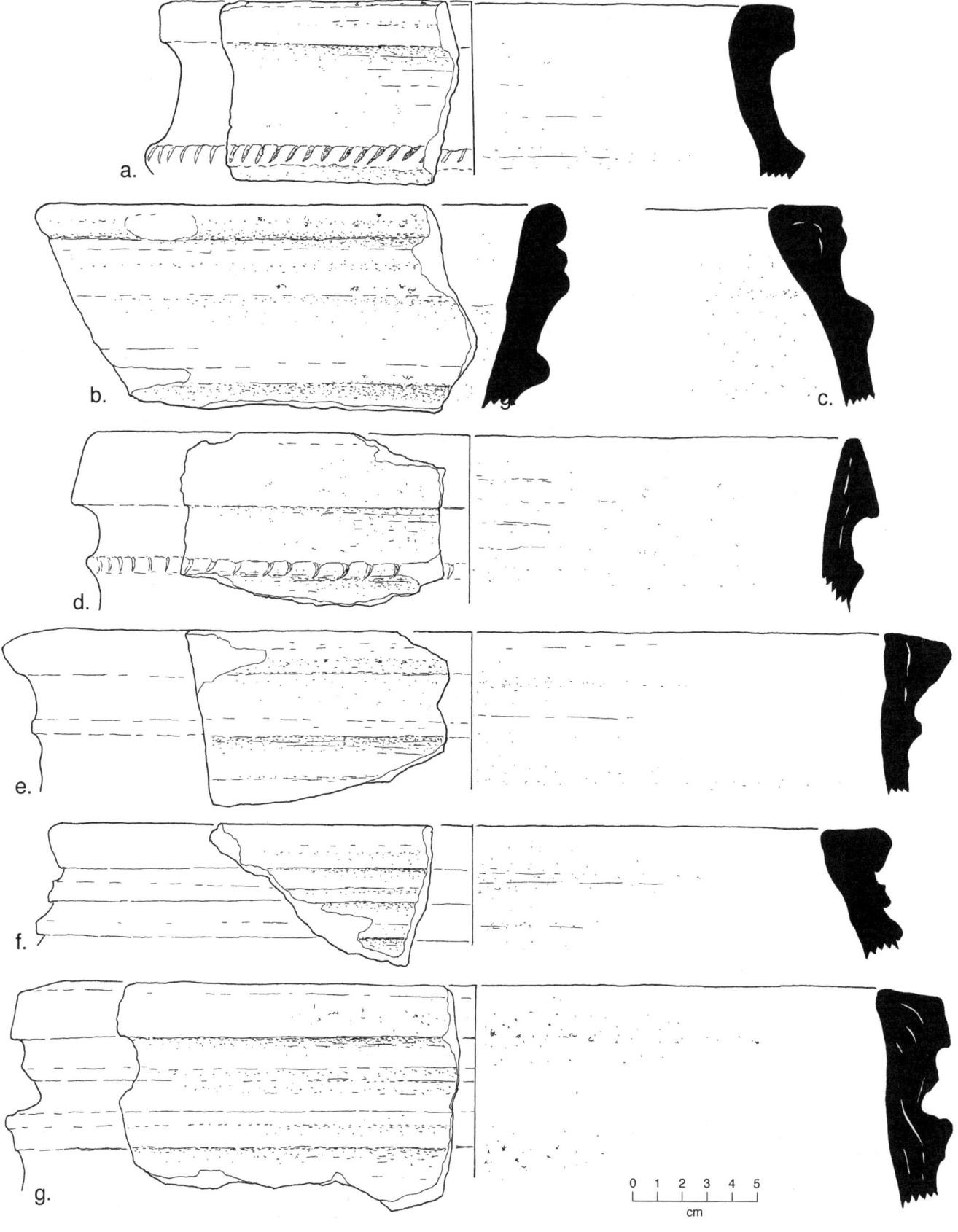

Figure 4.18. Basins or large jars of the IInd millennium B.C. from Tenel Ramon (DL-27).

The Archaeological Sites and Their Interpretation

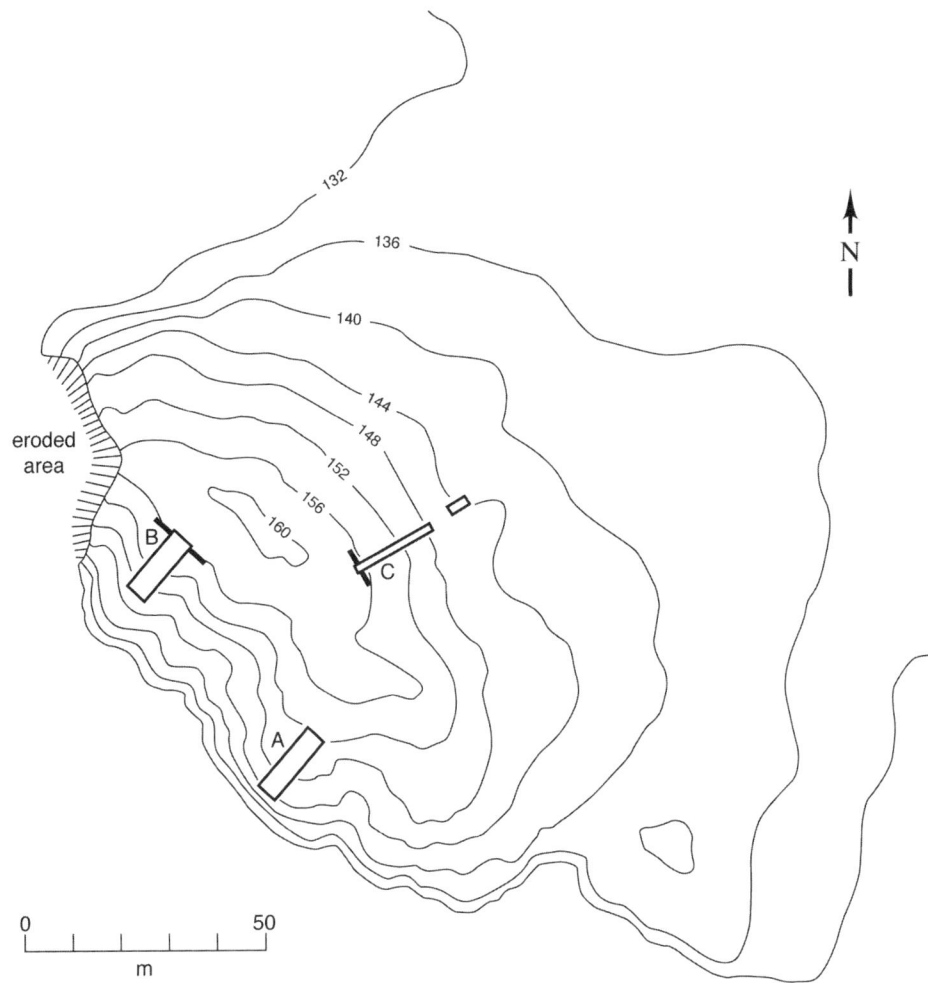

Figure 4.19. Map of Tepe Farukhābād (DL-32) showing Sukkalmaḫḫu revetment. A, B, and C are the 1968 excavations (Wright ed. 1981). The thick lines by B and C indicate the Sukkalmaḫḫu rampart.

Figure 4.18. Basins or large jars of the IInd millennium B.C. from Tenel Ramon (DL-27).

a, Basin with square rim (SM-RB-6) (Area 1.4). 5% vegetal inclusions, Dm 25, RT 2.59, ST 1.42, greenish-white (lip worn from use) (5Y 8/2) body (Farukhābād B13–16 [Sukkalmaḫḫu]: Fig. 85L, B11–13 [Sukkalmaḫḫu]: Fig. 86L).

b, Basin with grooved rim (SI-RB-4) (Area 1.4). 10% vegetal and fine sand inclusions, Dm 47, RT 1.57, ST 1.30, pale yellow (2.5Y 8/3) body (Farukhābād B19 [Šimaški]: Fig. 84h).

c, Basin with flat rim (SI-RB-5) (Area 2.1). 10% vegetal and medium sand inclusions, Dm ca. 55, RT 3.02, ST 1.18, pale yellow (2.5Y 8/3) body (Farukhābād B15–16 [Šimaški]: Fig. 90f, B12 [Sukkalmaḫḫu]: Fig. 86k).

d, Basin with band rim (SM-RB-7) (Area 2.1). 15% vegetal and medium sand inclusions, Dm 32, RT 1.97, ST 1.08, white (2.5Y 8/2) body (cf. Farukhābād B11–13 [Sukkalmaḫḫu]: Fig. 86L; Suse Ville Royale AXII [Sukkalmaḫḫu] Group 35b: Pl. 44:3, 4).

e, Basin with flat rim (SM-RB-4) (Area 2.1). 15% vegetal and medium sand inclusions, RT 2.53, ST 1.08, white (10YR 8/2) body (cf. Farukhābād B13 [Sukkalmaḫḫu]: Fig. 86n).

f, Basin with square rim (SM-RB-6) (Area 2.1). 10% vegetal inclusions, Dm 34, RT 2.85, ST 1.44, pink (5YR 7/4) body, white (10YR 8/2) surface (Farukhābād B15 [Šimaški]: Fig. 90h).

g, Basin with square rim (SM-RB-6) (Area 1.4). 10% vegetal and medium sand inclusions, Dm 37, RT 2.40, ST 1.55, RHt 2.35, pale yellow (2.5Y 8/3) body (Farukhābād B13–16 [Sukkalmaḫḫu-Šimaški]: Figs. 86i, 90h; Suse Ville Royale AXIV–XV [Sukkalmaḫḫu] Group 36a: Pl. 45:2–8).

Number: DL-34
Name: Tepe Gārān (also Guran, Goughan)
Grid: E 47°20'14" / N 32°34'43"
  NIOC: E 1719.1 – N 1177.4
  UTM: E 719,391 / N 3,606,984
Dimensions:
  Length: 480 m (NE-SW)
  Width: 450 m (NW-SE)
  Height: ca. 20 m
  Area: 17 ha

Modern Environmental Context: On the alluvial fan of the Dawairij or Āb-Dānān River, near the juncture with the northeastern alluvial slopes of the plain, about three km northwest of the present Dawairij channel.

Features: Tepe Gārān is a large site (Fig. 4.20, Plate 4), evidently the major center perhaps as early as the late IInd millennium B.C. and certainly during the mid- to late Ist millennium B.C. on the Deh Lurān Plain. We previously estimated the site covered about 40 ha, but recent satellite images show that the mounded area, not including the surrounding erosional fans, covers only 17 ha. It is estimated that the height of the large, conical-shaped mound, which is the tallest architectural feature of the site, was about 20 m.

The site is comprised of five large, irregularly shaped, platform-like structures that lie east, north, and west of the large, very high, conical-shaped mound situated near the southern boundary of the site. An irregularly-shaped open area, perhaps representing a large plaza or *meydan*, lies northeast of the high conical-shaped mound and separates it by a substantial distance from the northernmost large flat mound situated near the northern edge of the site. Linear gravel concentrations were visible in the west part of the site. These call to mind the gravel footings used for Achaemenid public buildings at Susa. Several smaller mounds are clearly visible atop the large platforms to the east and west of the large conical mound. Three small mounds lie south of the conical mound, and form the southernmost constructions of the site. About eight small mounds lay a short distance to the northwest of the Tepe Gārān complex of platforms and mounds. As a lack of time did not permit us to examine carefully these mounds, it is not known if they were part of this site complex. These have since been leveled.

Because of the magnitude of this site, and the time constraints placed on this survey, an accurate map of Tepe Gārān was not drafted in the field. The accompanying plan map of this site has been drafted from recent satellite images.

Phases: We have both a large 1968 general collection made by the Farukhābād team primarily from the high mound (Area 1) and a more systematic sector by sector collection made by Neely in 1969. Careful restudy of the complete Gārān collection has identified more sherds of the Early Dynastic period than were recognized in our first study (Neely and Wright 1994:102). We have recorded solid footed conical cups (ED-BC-1: Fig. 4.21*a*), narrow conical cup bases (ED-BC-2: Fig. 4.21*d*), wide conical cup bases (ED-BC-3: Fig. 4.21*c*), a beveled lip bowl (ED-RB-2: Fig. 4.21*h*), a round lip jar (ED-RJ-1: Fig. 4.21*i*), a ledge rim jar (ED-RJ-4: Fig. 4.21*j*), a flared expanded rim jar (ED-RJ-3) with monochrome decoration (ED-O-3: Fig. 4.21*k*), a sherd from the shoulder of a small carinated pot with a monochrome crosshatch motif (ED-O-5), a jar shoulder with a plain raised strip (ED-O-1), and a pinched ring jar base (ED-B-5). Most of these sherds are typical of the Earlier Early Dynastic I–II phase. Only a carinated sherd with monochrome design (not illustrated) is diagnostic of the Later Early Dynastic III phase. It is likely that occupation was reduced in the later period.

A few sherds indicate Šimaški phase occupation. The rim of a heavy concial cup with vegetal temper (SI-RC-1: Fig. 4.21*b*) is a Šimaški type common at Susa but could also be Neo-Elamite. The rims of indented rim carinated cups (SI-RC-3: Fig. 4.21*e*), a string-cut flat base with lip (SI-BC-1: Fig. 4.21*f*), and a carinated band rim jar with grooved shoulder (Fig. 4.22*h*) are well dated at Susa. A grooved band rim basin (SI-RB-4) with wavy incising (SI-O-2: Fig. 4.21*m*) is well dated at Farukhābād. These few sherds were found on the high mound (Area 1), and it is possible that the Šimaški occupation is buried deeply under later deposits on this mound.

The next attested occupation is during the Sukkalmaḫḫu and Middle Elamite periods. Some sherds are of types made in both Sukkalmaḫḫu and earlier Middle Elamite times. Among these are the sherds of goblets (SM-BG-4: Fig. 4.22*a*), a ledge rim bowl (SM-RB-3: Fig. 4.22*i*), a flat lip bowl (SM-RB-4: Fig. 4.22*g*), a ledge rim basin (SM-RB-5: Fig. 4.22*j*), a flat lip basin (SM-RB-8: Fig. 4.22*f*), and a band rim jar (SM-RJ-4: Fig. 4.22*d*), which are widely scattered across north and west portions of the site in Areas 1, 2, 3, and 5. Characteristic of the Sukkalmaḫḫu occupation is an oblique rim jar (SM-RJ-3: Fig. 4.22*b*), and basins with square rim (SM-RB-6: Fig. 4.22*k, l*), all from the high mound (Area 1). Indicative of Middle Elamite occupation are a tall goblet variant (SM-RG-2: Fig. 4.22*c*), probably from the high mound, and a flared slightly indented variant of the band rim jar (SM-RJ-5: Fig. 4.22*e*) from the north edge of the site. Both of these types, however, may prove to continue in Later Middle Elamite times when excavated IInd millennium samples are available from Deh Lurān. Surprisingly, there are no definite examples of the goblet base with ridge in the collections. However, Carter's (1971) suggestion that after the abandonment of Musiyān, at some time during the Middle Elamite Period, the town of Gārān grew substantially is plausible.

We have looked carefully in the Gārān collections for the early Ist millennium Neo-Elamite types defined by Miroschedji (1981b), and the sole likely example is the rim of a straight neck jar with beaded rim and neck ridge (NE-RJ-2: Fig. 4.24*l*) from the high mound. The possibility of a reestablishment of occupation on the high mound in Neo-Elamite times must be investigated in the future.

By the Achaemenid period, Gārān had clearly emerged as the major center on the Deh Lurān Plain. On all areas of the site, sherds of hard, sandy ware, often compacted or slightly

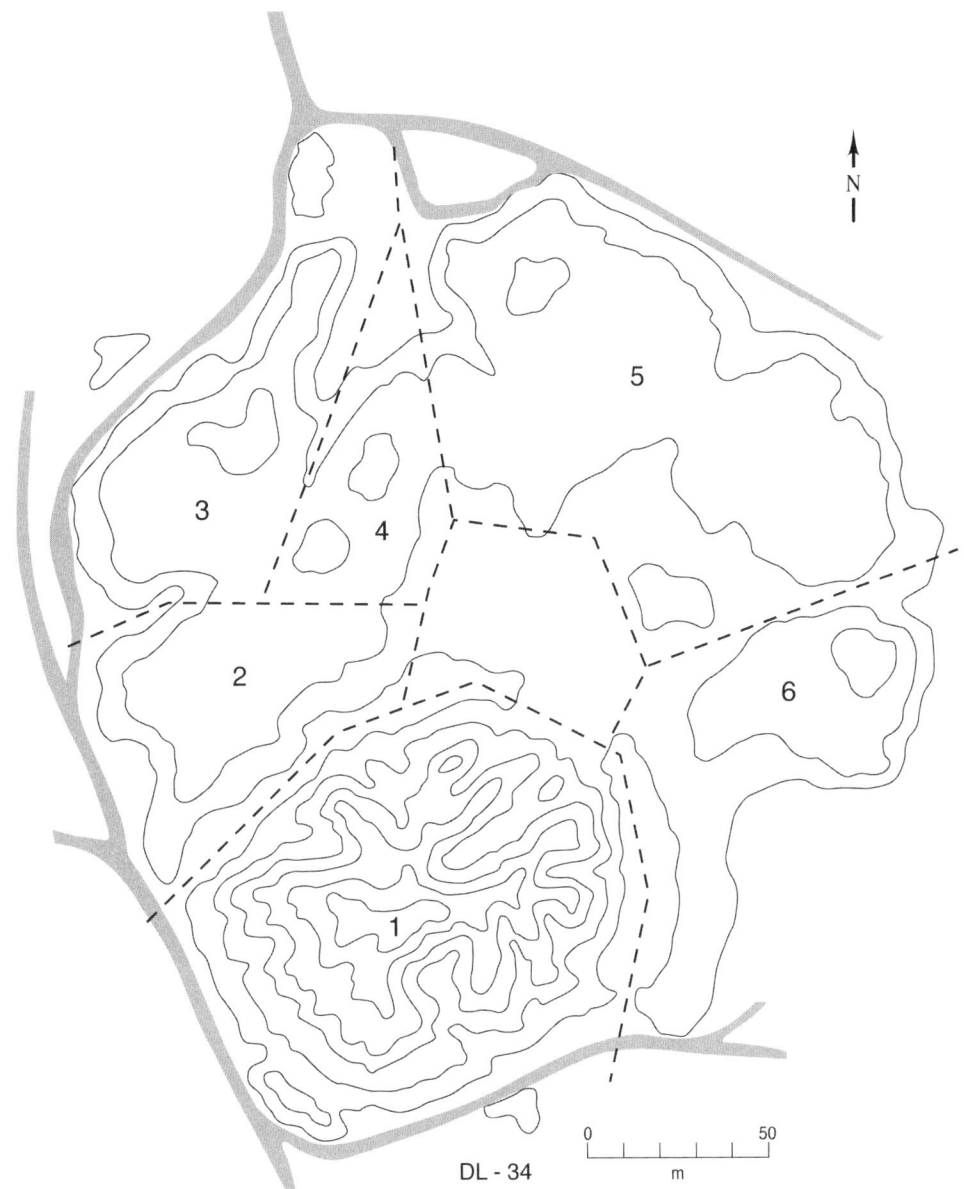

Figure 4.20. Map of Tepe Gārān (DL-34). (Gray lines mark recent canals.)

burnished, were found from carinated bowls, usually with beaded rims (AM-RB-2, 3: Fig. 4.23*b–h*), from a distinctive ledge rim bowl (AM-RB-4: Fig. 4.23*j, k*), from a basin with band rim (AM-RB-6: Fig. 4.23*a*), from neckless beaded rim jars (AM-RJ-2: Fig. 4.24*a, d*), and from flared beaded rim jars (AM-RJ-3: Fig. 4.24*c, e–k, m*). There are a few cups with a sandy fabric with a glaze, now oxidized to a whitish powder, among which we illustrate an indented rim variant (AM-RC-3: Fig. 4.23*i*) and a tooled ring base (AM-BC-2: Fig. 4.23*k*). Doubtless the excavation of Achaemenid contexts will allow us to recognize other types. The known examples are found in all parts of the site, which must have been fully occupied at this time.

Major occupation continued at Gārān in Hellenistic times and later. These periods will be discussed in our final volume.

Comment: The most conservative interpretation of the evidence of IIIrd millennium occupation at Tepe Gārān is that somewhere under the northern end of Area 6 is a site of the Early Dynastic period, to which we ascribe an area of 1.5 ha. For the Šimaški and Sukkalmaḫḫu occupations, we suggest the area of the high mound, about 1.8 ha. For the Middle Elamite expansion, we include the northern and western lower town as well as the high mound, a mounded area of about 12 ha. For the Achaemenid occupation, the full site area of about 17 ha must have been occupied. See Chapter 6 for a discussion of associated Canal DL-121.

References: Gautier and Lampre 1905:61; Carter 1971:231–34.

Figure 4.21. Cup, bowl, and jar sherds of the IIIrd millennium B.C. from Tepe Gārān (DL-34).

a, Solid footed goblet base (ED-BC-1) (Area 1x). 10% fine sand inclusions, oval base 3.0 × 2.8, pink (7.5YR 8/4) body, white (10YR 8/1) surface.

b, Wide conical cup (SI-RC-1) (Area 4). 5% vegetal inclusions, Dm 14, RT 0.57, ST 0.53, reddish-yellow (5YR 7/5) body, very pale brown (10YR 8/2) surface (Farukhābād A7 [EDI]: Fig. 45i).

c, Wide conical cup base (ED-BC-3) (Area 4). 5% fine sand inclusions, base Dm 3.8, ST 0.68, pale yellow (5Y 8/3) body (Farukhābād A6 [EDI+]: Fig. 45c).

d, Narrow conical cup base (ED-BC-2) (Area 3/2). 10% medium sand inclusions, base Dm 4.0, ST 0.75, pink (7.5YR 8/4) body, very pale brown (10YR 8/2) surface (Farukhābād AB23 [EDI]: Fig. 45d).

e, Indented rim carinated cup (SI-RC-3) (Area 1c). 15% medium sand and calcite inclusions, Dm 20, RT 0.53, ST 0.46, RHt 1.54, light reddish-brown (5YR 6/5) body, white (2.5Y 8/2) surface (Farukhābād B16–17 [Šimaški]: Fig. 84b; Suse Ville Royale BVII–VI [Šimaški] Group 6a: Pl. 7:2, 3, 6, 7, 10; Suse Ville Royale I, Level 6 [VA]: Fig. 38:3, Level 4 [VB]: Fig. 41:1, Level 3 [VB]: Figs. 45:4–6, 49:10, 12, 13).

f, Cup base (SI-BC-1) (Area 2). 10% medium sand inclusions, base Dm 4.9, ST 0.57, pale yellow (5Y 8/3) body.

g, Indented rim carinated cup (cf. SI-RC-3) (Area 1x). 10% fine sand and calcite inclusions, Dm 18, RT 0.42, ST 0.37, RHt 1.3, light pinkish-gray (7.5YR 6/3) body, white (2.5Y 8/2) surface (while the rim is indistinguishable from Late IIIrd millennium forms [same as comparanda e] the depth of this example is similar to Parthian forms, cf. Suse Apadana Fig. 67:1).

h, Beveled lip bowl (ED-RB-2) (Area 2). 10% fine sand inclusions, Dm 38, RT 1.39, ST 0.83, RHt 1.09, color not recorded (Farukhābād B31 [Late Uruk]: Fig. 47d, B25 [Jemdet Nasr]: Fig. 47j).

i, Flared round rim jar (ED-RJ-1) (Area 6). 5% medium sand inclusions, Dm 13, RT 0.67, NT 0.67, ST 0.63, pink (5YR 7/4) body, white (2.5Y 8/2) surface (Farukhābād A3–7 [EDI–II+]: Fig. 49b, k, L).

j, Ledge rim jar (ED-RJ-4) (Area 4). 5% vegetal and fine sand inclusions, Dm 14, RT 1.34, NT 0.67, RHt 1.19, greenish-white (5Y 8/2) body (Farukhābād A13 [Jemdet Nasr]: Fig. 53f–h, B24 [Jemdet Nasr]: Fig. 53n, B22 [EDI]: Fig. 53q).

k, Flared expanded rim jar (ED-RJ-3) with monochrome bands (ED-O-3) (Area 6). Trace 5% fine sand and vegetal inclusions, Dm 17, RT 0.78, NT 0.50, RHt 0.56, very pale brown (10YR 8/4) body, white (2.5Y 8/2) surface, reddish-brown (2.5YR 4/3) painted bands (Farukhābād B22 [EDI–II]: Fig. 56m).

l, Oblique rim jar (Area 4). 20% medium sand inclusions, Dm 22, RT 1.55, NT 0.79, RHt 1.04, pink (5YR 7/4) body, pinkish-white (7.5YR 8/2) surface (Farukhābād B22 [EDI+]: Fig. 53q).

m, Grooved band rim basin (SI-RB-4) with wavy incising (SI-O-2) (Area 1c). 15% medium sand inclusions, Dm 42, RT 1.34, NT 1.10, RHt 3.08, greenish-white (5Y 8/2) body (Farukhābād B17–19 [Šimaški]: Fig. 84h, j).

Figure 4.21. Cup, bowl, and jar sherds of the IIIrd millennium B.C. from Tepe Gārān (DL-34).

Figure 4.22. Goblets, jars, bowls, and basins of the IInd millennium B.C. from Tepe Gārān (DL-34).

a, Button base of goblet (SM-BG-4) (Area 5). 5% vegetal inclusions, base Dm 2.7, ST 0.72, pink (7.5YR 7/4) body, white (2.5Y 8/2) surface (Farukhābād B11–13 [Sukkalmaḫḫu]: Fig. 87*b*, *c*, *e*, B7 [Early Middle Elamite]: Fig. 89*r*).

b, Oblique rim jar (SM-RJ-3) (Area 1c). 5% vegetal and fine sand inclusions, Dm 18, RT 1.56, NT 0.92, RHt 1.81, pale yellow (5Y 7/3) body (Farukhābād B15–17 [Šimaški]: Fig. 84*m*, B11–13 [Sukkalmaḫḫu]: Fig. 86*b*).

c, Tall goblet (SM-RG-2) with button base (SK-BG-4) (1968 General Collection). 15% fine sand and vegetal inclusions, neck Dm 5.7, body Dm 8.4, base Dm 3.6, NT 0.66, ST 0.86, pale yellow (5Y 8/3) body (Suse Ville Royale AXI [Earlier Middle Elamite] Group 19a: Pl. 19:1, 3, 4; the puzzling attribute of this goblet is the predominance of sand inclusions, but these are known from other Early Middle Elamite goblets from the Deh Lurān area [cf. Fig. IV.6*i*], as well as from later "classical" Middle Elamite goblets [Miroschedji 1981: Fig. 12:15]).

d, Band rim jar (SM-RJ-4) (Area 2). 10% vegetal inclusions, Dm 14, RT 1.10, NT 0.72, RHt 1.48, white (2.5Y 8/2) body (Farukhābād B11–13 [Sukkalmaḫḫu]: Fig. 86*b*).

e, Indented band rim jar (SM-RJ-5) (Area 0). 10% vegetal and fine sand inclusions, Dm 14, RT 1.18, NT 0.61, RHt 2.32, white (2.5Y 8/2) body (Farukhābād B1–3 [Early Middle Elamite]: Fig. 89*i*; Suse Ville Royale AIX–XI [Earlier Middle Elamite] Group 17a: Pl. 18:3, 4; Suse Ville Royale II, Level 10 [Later Middle Elamite]: Fig.13:13).

f, Large basin with flat lip (SM-RB-8) (Area 3). 10% vegetal and coarse sand inclusions, Dm 36, RT 3.67, ST 1.47, white (2.5Y 8/2) body (Farukhābād B6 [Early Middle Elamite]: Fig. 89*q*; Suse Ville Royale AXV [Sukkalmaḫḫu] Group 36a: Pl. 45:5; Suse Ville Royale I, Level 3, 4A [VB]: Fig. 51:5).

g, Bowl with flat lip (SM-RB-4) (Area 5). 5% vegetal inclusions, Dm 26, RT 2.61, ST 0.95, RHt 1.56, white (2.5Y 8/2) body (Farukhābād B11–13 [Sukkalmaḫḫu]: Fig. 86*m*).

h, Carinated band rim jar with grooved body (Area 1c). 10% vegetal and fine sand inclusions, Dm 13, RT 1.44, NT 0.69, RHt 1.72, pinkish gray (10YR 6/3) body, white (2.5Y 8/1) surface (Suse Ville Royale BVI [Šimaški] Group 16: Pl. 17:6, 7).

i, Bowl with ledge rim (SM-RB-3) (Area 1x). 10% vegetal and fine sand inclusions, Dm 24, RT 2.29, ST 0.94, RHt 2.29, greenish-white (5Y 8/2) body (Farukhābād B12 [Sukkalmaḫḫu]: Fig. 85*k*, *o*; Suse Ville Royale I, Level 3 [VB]: Fig. 49:19).

j, Basin with ledge rim (SM-RB-5) with wavy incising (Area 4). 5% vegetal inclusions, Dm 34, RT 2.48, ST 1.34, RHt 2.21, white (2.5Y 8/2) body (Farukhābād B13 [Sukkalmaḫḫu]: Fig. 86*j*).

k, Basin with square rim (SM-RB-6) (Area 1c). 5% vegetal inclusions, Dm 27, RT 2.68, ST 1.33, RHt 2.47, white (5Y 8/1) body (Farukhābād B13–17 [Šimaški-Sukkalmaḫḫu]: Fig. 86*c*, *i*, *L*).

l, Basin with square rim (SM-RB-6) (Area 1c). 5% vegetal inclusions, Dm ca. 70, RT 3.17, ST 1.95, RHt 3.95, white (2.5Y 8/2) body (Suse Ville Royale AXI–XIV [Sukkalmaḫḫu] Group 33bc: Pl. 42:1–8).

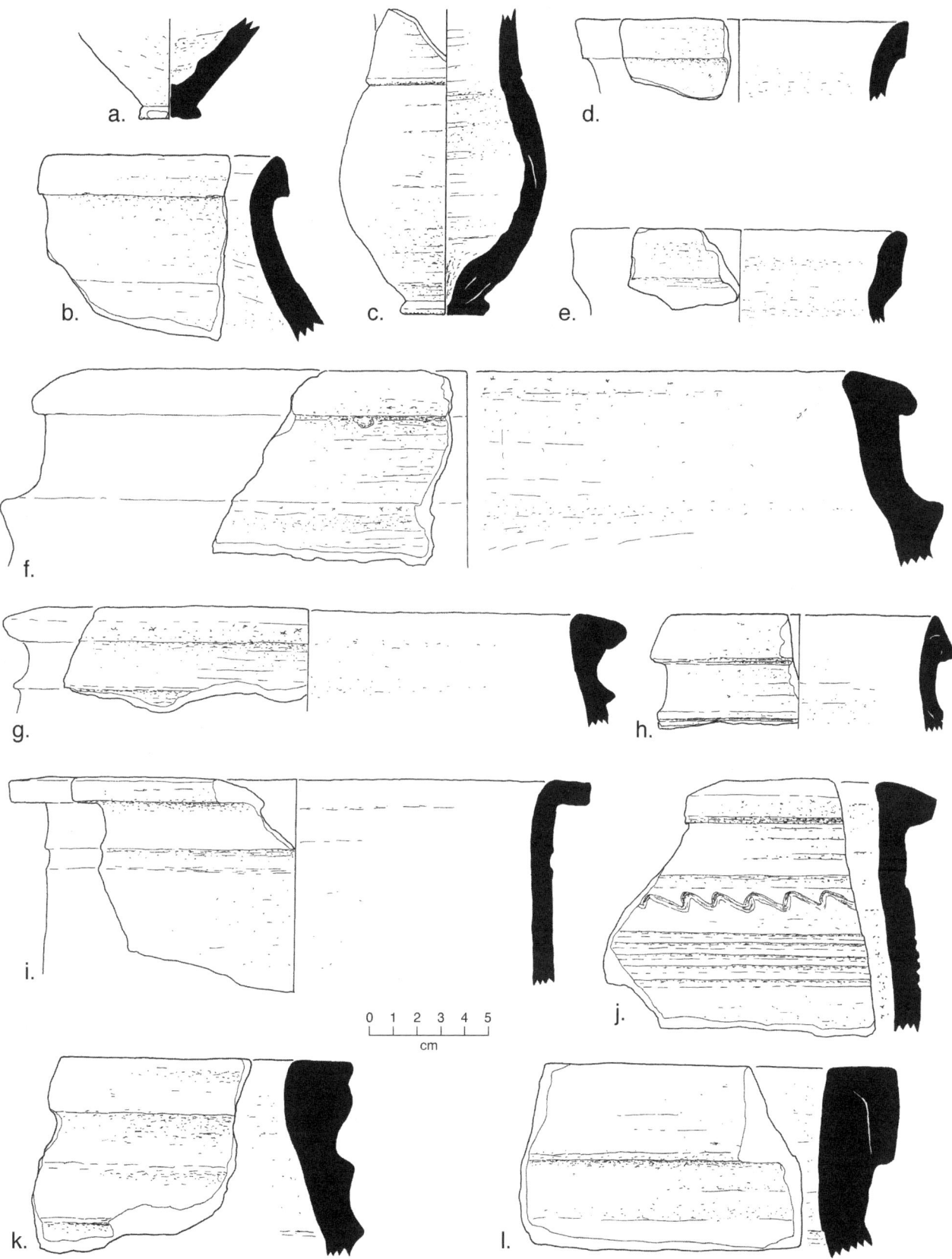

Figure 4.22. Goblets, jars, bowls, and basins of the IInd millennium B.C. from Tepe Gārān (DL-34).

Figure 4.23. Basin and bowls of the mid-Ist millennium B.C. from Tepe Gārān (DL-34).

a, Basin with band rim (AM-RB-6) (Area 0 [north and west periphery]). 15% medium sand and vegetal inclusions, Dm ca. 33, RT 1.75, ST 1.32, RHt 3.71, pale yellow (5Y 8/3) body, over-fired and warped (Suse Ville Royale II, Level 5A [Achaemenid]: Fig. 11:4).

b, Carinated bowl with curved rim (Area 5). 15% fine sand and fine calcite inclusions, Dm 31, RT 1.49, ST 1.01, RHt 1.08, color not recorded.

c, Carinated bowl with beaded rim (AM-RB-2) (Area 0). 10% medium sand inclusions, Dm ca. 24, RT 1.21, ST 0.74, RHt 0.95, reddish-yellow (5YR 6/5) body, very pale brown (10YR 8/3) surface (Suse Ville Royale II, Level 5B [Achaemenid]: Fig. 9:1).

d, Carinated bowl with beaded rim (AM-RB-2) (Area 3). 5% fine sand and calcite inclusions, Dm 38, RT 1.33, ST 0.75, RHt 1.74, light brownish-gray (10YR 6/2) body, light gray (2.5Y 7/2) surface.

e, Carinated bowl with beaded rim (AM-RB-2) (Area 3). 10% fine sand, Dm 23, RT 1.42, ST 0.43, RHt 1.33, pinkish gray (7.5YR 7/3) body (Suse Ville Royale II, Level 5B [Achaemenid]: Fig. 9:1).

f, Carinated bowl with beaded rim (AM-RB-2) (Area 3). 10% fine sand, Dm 21, RT 1.19, ST 0.64, RHt 0.96, light reddish brown (5YR 6/4) body, very pale brown (10YR 7/3) surface (Suse Ville Royale II, Level 5 [Achaemenid]: Fig. 7:11).

g, Carinated bowl with beaded rim (AM-RB-2) (Area 6). 10% fine calcite, Dm 37, RT 1.40, ST 0.63, RHt 1.33, light brown (7.5YR 6/4) body, white (10YR 8/2) surface, pinkish gray (7.5YR 6/2) painted bands.

h, Deep carinated bowl with beaded rim (AM-RB-3) (Area 3). 5% fine sand and calcite, Dm 33, RT 1.35, ST 0.41, RHt 1.51, light brown (7.5YR 6/4) body, very pale brown (10YR 7/3) slip.

i, Indented band rim bowl (AM-RC-3) (1968 General Collection). No visible inclusions, Dm 21, RT 0.61, ST 0.33, RHt 1.75, very pale brown (10YR 8/3) body, powdery white glaze remnant (cf. Suse Ville Royale II, Level 4 [Achaemenid]: Fig. 9:10).

j, Carinated bowl with ledge rim (AM-RB-4) (Area 5). Trace fine sand vegetal inclusions, Dm 21, RT 1.95, ST 0.58, RHt 0.57, very pale brown (10YR 7/3) body, light reddish-brown (10YR 2/5) surface, burnished exterior surface (Suse Ville Royale II, Level 5B [Achaemenid]: Fig. 7:12; Apadana-Ville Royal 6B [Achaemenid]: Fig. 57:1).

k, Carinated bowl with ledge rim (AM-RB-4) (Area 3). 5% fine sand inclusions, Dm 27, RT 2.60, ST 0.62, RHt 0.89, pale brown (10YR 6/3) body, light reddish brown (5YR 6/5) surface (same camparanda).

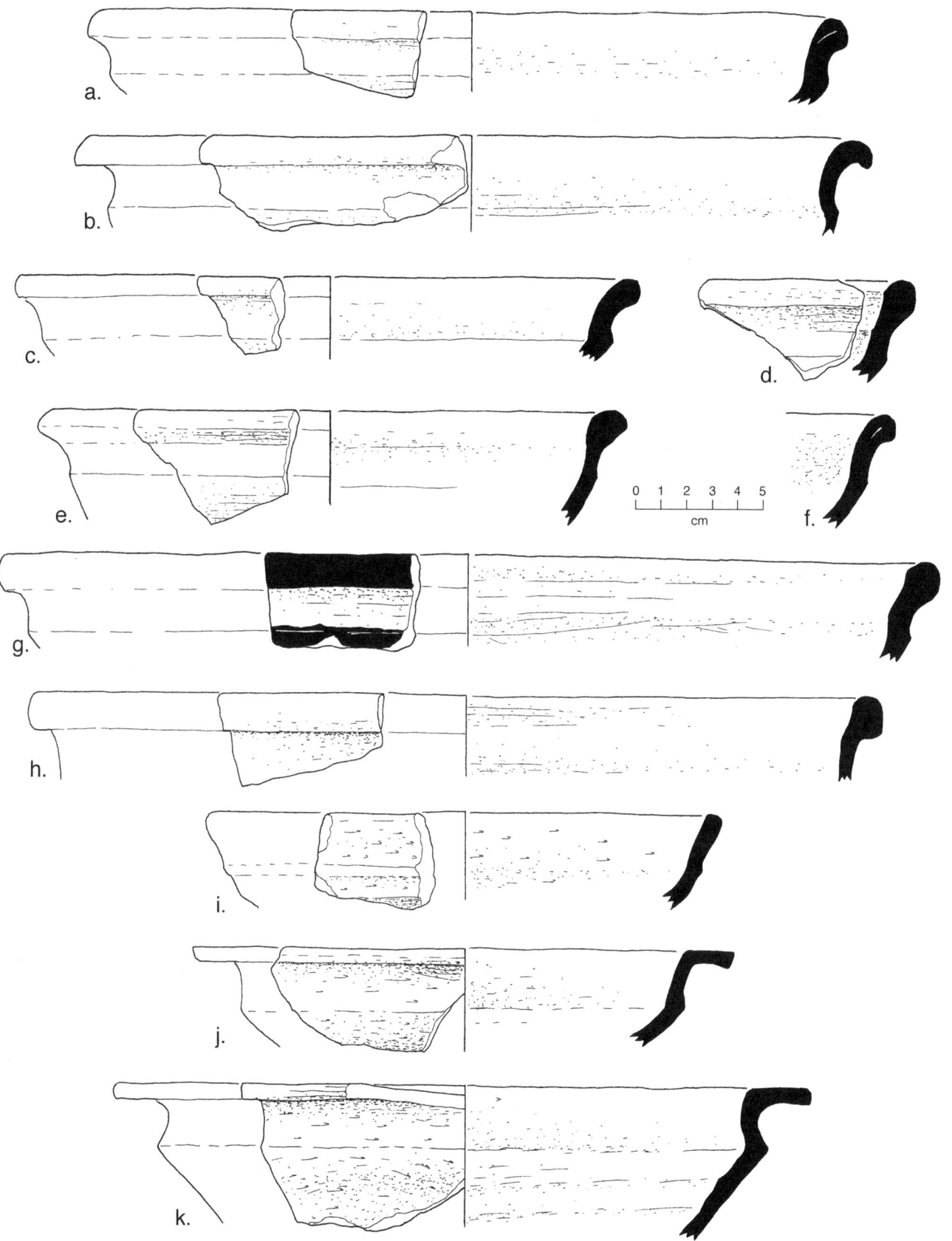

Figure 4.23. Basin and bowls of the mid-Ist millennium B.C. from Tepe Gārān (DL-34).

Figure 4.24. Jars of the mid-Ist millennium B.C. from Tepe Gārān (DL-34).

a, Neckless jar with beaded rim (AM-RJ-2) (Area 0). 15% fine sand inclusions, Dm 25, RT 1.92, NT 0.89, RHt 1.69, pink (7.5YR 6/5) body, surface eroded.

b, Flared neck jar with ledge rim (Area 2). Trace fine sand inclusions, Dm 22, RT 1.51, NT 0.78, RHt 1.21, light reddish-brown (5YR 6/4) body (Suse Ville Royale II, Level 5A [Achaemenid]: Fig. 12:4).

c, Flared neck jar with beaded rim (AM-RJ-3) (Area 3). 10% medium sand inclusions, Dm 13, NT 0.75, RT 1.37, ST 0.75, RHt 1.33, white (2.5Y 8/2) body (Suse Ville Royale II, Level 4 [Achaemenid]: Fig. 15:2, 10).

d, Neckless jar with beaded rim (1968 General Collection). 5% medium sand and vegetal inclusions, Dm 17, RT 1.40, NT 0.87, ST 0.99, RHt 1.72, pinkish gray (7.5YR 6/2) body, very pale brown (10YR 8/3) surface (Suse Ville Royale II, Level 4 [Achaemenid]: Fig. 15:5, 9).

e, Flared neck jar with beaded rim (AM-RJ-3) (Area 1c). 5% fine sand and vegetal inclusions, Dm 14, RT 1.30, NT 0.95, RHt 1.20, greenish-white (5Y 8/2) body (Suse Ville Royale II, Level 11–12 [Later Middle Elamite]: Fig. 13:6, 7, Level 5A [Achaemenid]: Fig. 15:3).

f, Flared neck jar with expanded rim (AM-RJ-3) (Area 6). Trace of fine sand inclusions, Dm 13, RT 0.81, NT 0.94, RHt 0.84, very pale brown (10YR 8/3) body.

g, Flared neck jar with beaded rim (AM-RJ-3) (Area 3). 10% fine sand and calcite inclusions, Dm 13, RT 1.09, NT 0.60, RHt 1.24, light brown (7.5YR 6/4) body, pinkish white (7.5YR 8/3) surface (Suse Ville Royale II, Level 5A [Achaemenid]: Fig. 15:3).

h, Flared neck jar with beaded rim (AM-RJ-3) (Area 0). 15% fine sand inclusions, Dm 17, RT 1.70, NT 0.88, RHt 1.42, light reddish-brown (5YR 6/4) body, white (2.5Y 8/2) surface (Suse Ville Royale II, Level 4 [Achaemenid]: Fig. 15:5).

i, Flared neck jar with beaded rim (AM-RJ-3) (Area 1x). 5% fine sand and calcite inclusions, Dm 12, RT 1.09, NT 0.62, RHt 1.20, pink (7.5YR 7/4) body, white (2.5Y 8/2) surface (Suse Ville Royale II, Level 5A [Achaemenid]: Fig. 16:6, 8).

j, Flared neck jar with beaded rim (AM-RJ-3) (Area 0). Trace vegetal and fine sand inclusions, Dm ca. 37, RT 1.62, NT 0.78, RHt 1.50, pink (5YR 7/4) body, very pale brown (10YR 8/3) surface, exterior burnish.

k, Flared neck jar with beaded rim (AM-RJ-3) (Area 0). 5% fine sand and vegetal inclusions, Dm ca. 30, RT 1.53, NT 0.80, RHt 1.26, reddish yellow (7.5YR 6/5) body, pink (7.5YR 7/4) surface, interior slip.

l, Straight neck jar with beaded rim and neck ridge (NE-RJ-2) (Area 1x). Trace of fine calcite and fine sand inclusions, Dm 7, RT 0.64, NT 0.51, RHt 0.91, pink (7.5YR 8/4) body, very pale brown (10YR 8/3) surface, exterior burnish (Suse Ville Royale II, Level 9 [Neo-Elamite]: Fig. 23:10, cf. Level 5B [Achaemenid]: Fig. 16:2, 3).

m, Flared neck jar with beaded rim (AM-RJ-3) (Area 5). 10% fine sand and calcite inclusions, Dm 30, RT 1.68, NT 0.58, RHt 1.69, color not recorded (Suse Ville Royale II, Level 5A [Achaemenid]: Fig. 15:6).

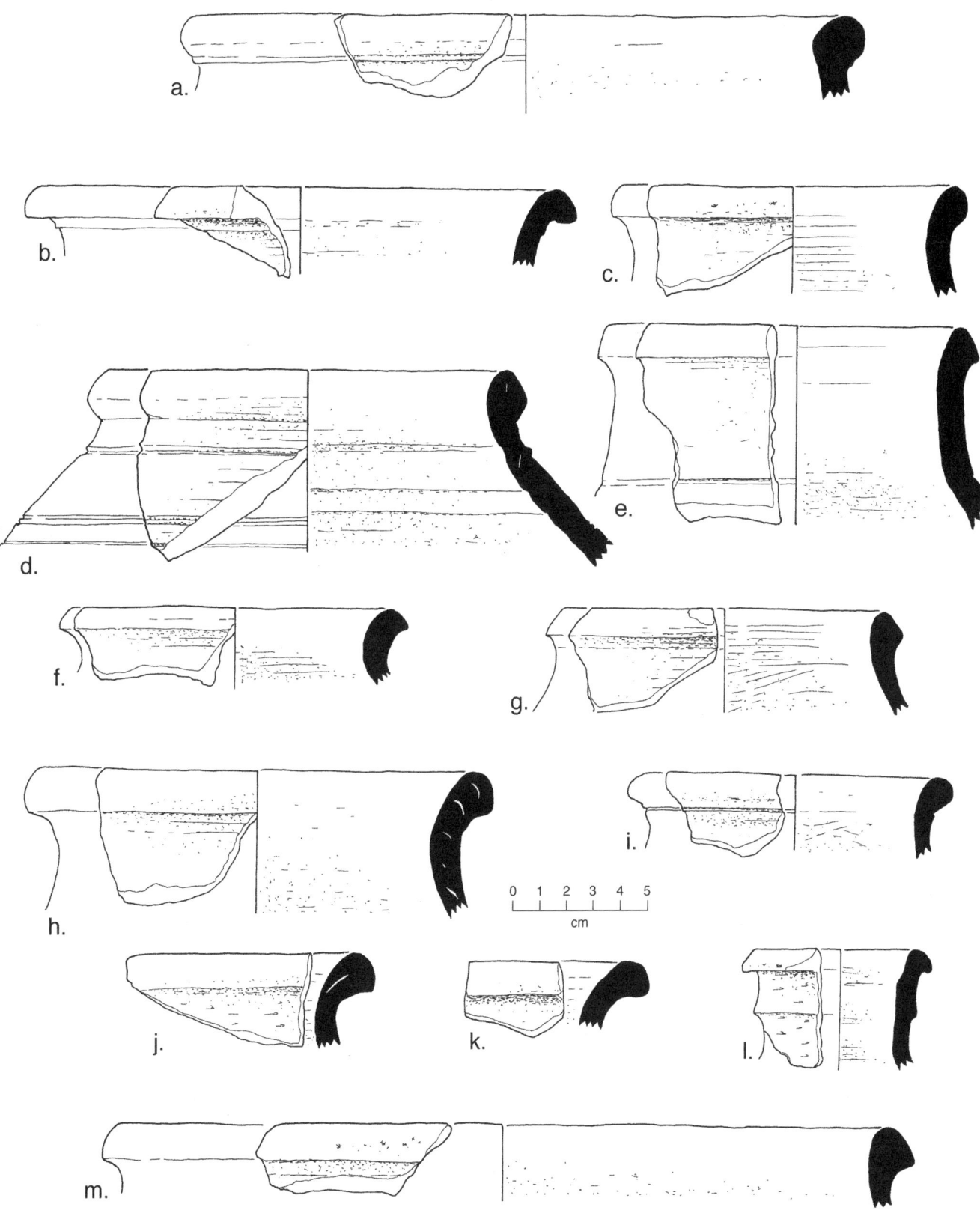

Figure 4.24. Jars of the mid-Ist millennium B.C. from Tepe Gārān (DL-34).

Number: DL-35 (Miroschedji DK-25)
Name: Tepe Patak
Grid: E 47°29'09" / N 32°29'47"
  NIOC: E 1733.3 – N 1172.9
  UTM: E 733, 558 / N 3, 598, 182
Dimensions:
  Length: 265 m (N-S)
  Width: 200 m (E-W)
  Height: 13.00 m
  Area: 6.5 ha

Modern Environmental Context: On gravel fans north of the Dawairij or Āb-Dānān River, about 3.5 km north of the present channel of the river, on the northeastern slopes of the plain.

Features: This is a rectangular site defined by traces of mud brick walls (Fig. 4.25, Plate 5). All areas of the site have early to mid-IInd millennium and mid-Ist millennium occupation, and it is likely that the site was walled during both periods.

Phases: The 1968 survey team made an extensive sectorial collection on this site. Pierre de Miroschedji also visited the site and discussed it in detail (Miroschedji 1981b).

Restudy of the Patak ceramics revealed that we had failed to record the full range of IIIrd millennium types in our earlier study (Neely and Wright 1994:104). Representing the Earlier Early Dynastic I–II phase are fragments of solid footed conical cup bases (ED-BC-1), a narrow conical cup base (ED-BC-2), wide conical cup bases (ED-BC-1), conical cup rims (ED-RC-1), and a beveled lip bowl (ED-RB-2). These are scattered in all areas of the site, but most are small and battered and may be in fill displaced from a relatively small ED I–II site. They could perhaps be items in mud bricks made at DL-41 and transported down Canal-317 by barge. A small jar shoulder (probably a fragment of ED-RJ-6) with painted monochrome design (ED-O-5) is probably from the Later Early Dynastic III phase. We have no grounds for estimating the size of the Early Dynastic occupations, but they were probably small.

Definite Šimaški basin rims (SI-RB-4: Fig. 4.27*e, f*) were scattered in the center of the site. All sectors of the site produced Sukkalmaḫḫu and earlier Middle Elamite materials including goblet rims (SM-RG-1, SM-RG-2: Fig. 4.26*a, c*) and button bases (SM-BG-4: Fig. 4.26*c, e*), jars with ledge rims (SM-RJ-1: Fig. 4.26*f*), band rims (SM-RJ-4: Fig. 4.26*k*) and square rims (SM-RJ-2: Fig. 4.26*j*), and basins with ledge rims (SM-RB-5: Fig. 4.27*c, d*). Characteristic of the Sukkalmaḫḫu phase are oblique rim jars (SM-RJ-3: Fig. 4.26*g, h*) and basins with band rims (SM-RJ-7: Fig. 4.27*a, b*). In contrast, a number of items are characteristically earlier Middle Elamite, among them goblet bases with a ridge (SM-BG-2: Fig. 4.26*b*) and an internal plug (SM-BG-3: Fig. 4.26*d*), and jars with indented band rims (SM-RJ-5: Fig. 4.26*l, m*). It is to this period that we can ascribe moldmade terra cotta objects representing beds with a male and female figure embracing. Both were broken in half so that only the upper bodies remain (Plate 8). Similar terra cottas were found at Susa in Ville Royale AXII and XIII, which date to the earlier Middle Elamite period.

Miroschedji (1981b) has made a cogent argument for an early Ist millennium Neo-Elamite occupation at Patak and reported a low wide mouthed goblet with neck ridge (NE-RG-1) (Miroschedji 1981b: Fig. 64.3) diagnostic of this time. We looked carefully for additional ceramic evidence, but the sole possible evidence in our collection is the rim of a flared neck jar with beaded rim and neck ridge (AM-RJ-5: Fig. 4.28*o*). This could be either Neo-Elamite or Achaemenid. The possibility of an occupation at Patak in Neo-Elamite times must be further investigated in the future.

By the Achaemenid period, Patak had clearly reemerged as a significant settlement on the Deh Lurān Plain. On all areas of the site were sherds of hard, sandy ware, often compacted or slightly burnished, from carinated bowls, usually with beaded rims (AM-RB-2: Fig. 4.28*d–k*), and from flared jars with beaded rims (AM-RJ-3: Fig. 4.28*l, m*). There were no glazed cups, but there is a range of incurved bowls with a light burnish (Fig. 4.28*a–c*), which resemble slipped Hellenistic types, and may indicate a particularly late Achaemenid or Post-Achaemenid occupation here.

Comment: We arbitrarily ascribe an area of 1.0 ha to Early Dynastic and Šimaški Patak. It appears that as Sukkalmaḫḫu Musiyān (DL-20) was diminishing, DL-35 was becoming the largest settlement of the eastern plain, covering about 5 ha. The archaeological evidence for Neo-Elamite occupation is limited, but by later Achaemenid times, the entire site area of 6.5 ha was occupied. See Chapter 6 for a discussion of Canal 317C.

Possible later occupations will be discussed in our final Deh Lurān survey volume.

References: Miroschedji 1981b:170, 174, 175, Fig. 64; Carter 1971:234–35.

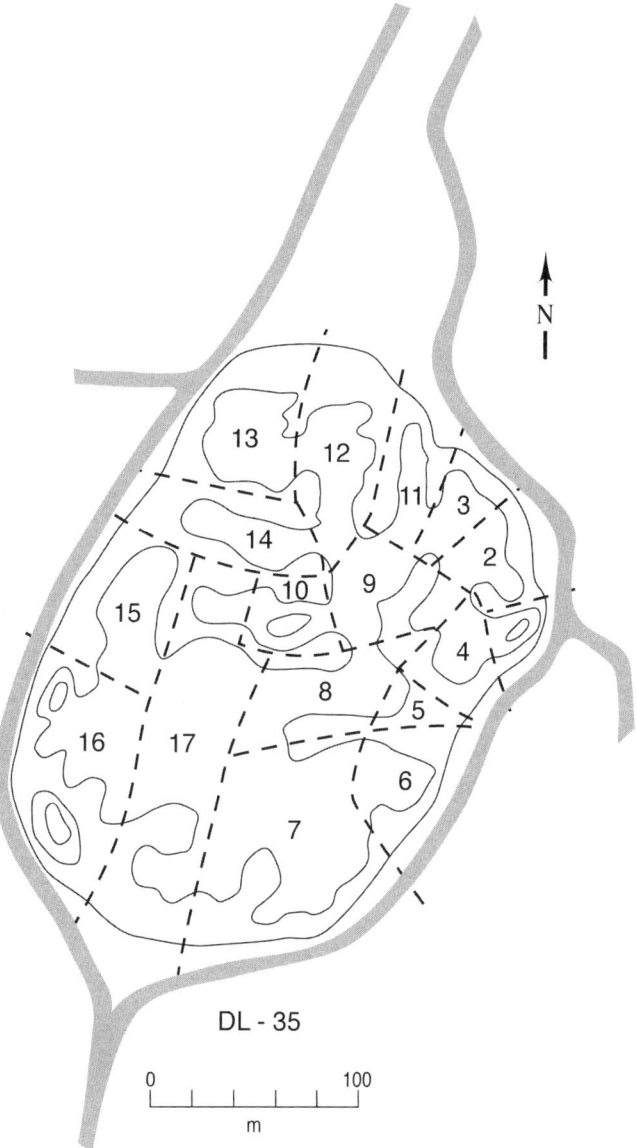

Figure 4.25. Map of Tepe Patak (DL-35). (Gray lines mark recent canals.)

Figure 4.26. Goblet, bowl, and jar sherds from Tepe Patak (DL-35).

a, High goblet rim (SM-RG-2) (Area 6–7). 5% vegetal inclusions, Dm 7, RT 0.27, NT 0.42, pink (7.5YR 7/4) body, very pale brown (10YR 8/3) surface (Farukhābād B18 [Šimaški]: Fig. 90g, B12–13 [Sukkalmaḫḫu]: Fig. 85f, g).

b, Goblet base with ridge (SM-BG-2) (Area 2). 5% medium sand and vegetal inclusions, base Dm 5.50, ST 0.72, pink (5YR 7/4) body, white (2.5YR 8/2) surface (Farukhābād B1–3 [Earlier Middle Elamite]: Fig. 89u).

c, Goblet (SM-RG-1) with button base (SM-BG-4) (Area 12). 5% vegetal and medium sand inclusions, rim Dm 8, body Dm 11.3, base Dm 2.70, goblet height 12.6, neck height 3.1, RT 0.30, NT ca. 0.40, red (5YR 6/5) body, very pale brown (10YR 7/3) surface (Farukhābād B12–13 [Sukkalmaḫḫu]: Fig. 90c; Suse Ville Royale AXI–XIII [Sukkalmaḫḫu] Group 20ab: Pl. 21:7, 9, 12, 20–23, 28, 29).

d, Goblet base with internal plug (SM-BG-3) (Area 12). 5% vegetal and fine sand inclusions, base 4.40 × 4.10, ST 0.59, pale yellow (5Y 8/3) body (Farukhābād B12 [Sukkalmaḫḫu]: Fig. 87b, c, B4 [Earlier Middle Elamite]: Fig. 89s; Suse Ville Royale AIX–XI [Earlier Middle Elamite] Group 19b: Pl. 19:6, 7, 9–15, 17–22).

e, Button base of goblet (SM-BG-4) (Area 6–7). 5% vegetal inclusions, Dm 3.5, ST 0.79, white (2.5YR 8/2) body (Farukhābād B11–17 [Šimaški-Sukkalmaḫḫu]: Fig. 85e, f, B4–7 [Earlier Middle Elamite]: Fig. 89r, s).

f, Bowl or jar with ledge rim (Area 6–7). 5% vegetal and medium sand inclusions, rim Dm 18, RT 1.43, NT 0.81, RHt 1.12, white (2.5Y 8/2) body (Farukhābād B15–16 [Šimaški]: Fig. 86e, B11–12 [Sukkalmaḫḫu]: Figs. 85k, 86a).

g, Jar with oblique rim (SM-RJ-3) (Area 2). 5% vegetal inclusions, Dm 17, RT 1.28, NT 0.68, RHt 1.66, greenish-white (5Y 8/2) body (Farukhābād B11–13 [Sukkalmaḫḫu]: Fig. 86b).

h, Jar with oblique rim (SM-RJ-3) (Area 6–7). 10% vegetal inclusions, Dm 28, RT 1.06, NT 0.57, RHt 1.19, white (2.5Y 8/2) body (Farukhābād B4 [Earlier Middle Elamite]: Fig. 89L).

i, Female figurine (Area 2). 10% medium sand inclusions, hip width 3.88, thickness 1.57, white (2.5Y 8/2) body (cf. Gautier and Lampre 1905: Fig. 120 [Musiyān]).

j, Jar with square rim (SM-RJ-2) (Area 2). 10% vegetal inclusions, Dm 16, RT 1.88, RHt 1.40, NT 0.85, pale yellow (5YR 8/3) body (Farukhābād B13–16 [Šimaški-Sukkalmaḫḫu]: Fig. 86i).

k, Jar with band rim (SM-RJ-4) (Area 6–7). 5% vegetal inclusions, Dm 16, RT 1.16, NT 0.68, RHt 2.20, light brown (7.5YR 6/4) body, organic coating on interior and exterior (cf. Farukhābād B15–17 [Šimaški]: Fig. 84m, B11–13 [Sukkalmaḫḫu]: Fig. 86b; Suse Ville Royale AXII [Sukkalmaḫḫu] Group 30a: Pl. 37:1).

l, Jar with indented band rim (SM-RJ-5) (Area 7). 5% vegetal inclusions, Dm 16, RT 0.89, NT 0.69, RHt 2.20, greenish-white (5Y 8/2) body (Farukhābād B1–3 [Earlier Middle Elamite]: Fig. 89i; Suse Ville Royale AIX–XI [Earlier Middle Elamite] Group 17a: Pl. 18:3, 4; Suse Ville Royale II, Level 10 [Later Middle Elamite]: Fig. 13:10).

m, Jar with indented band rim (SM-RJ-5) (Area 2). 10% vegetal and medium sand inclusions, Dm 16, RT 0.91, NT 0.72, RHt 2.28, white (2.5Y 8/2) body (same comparanda as l).

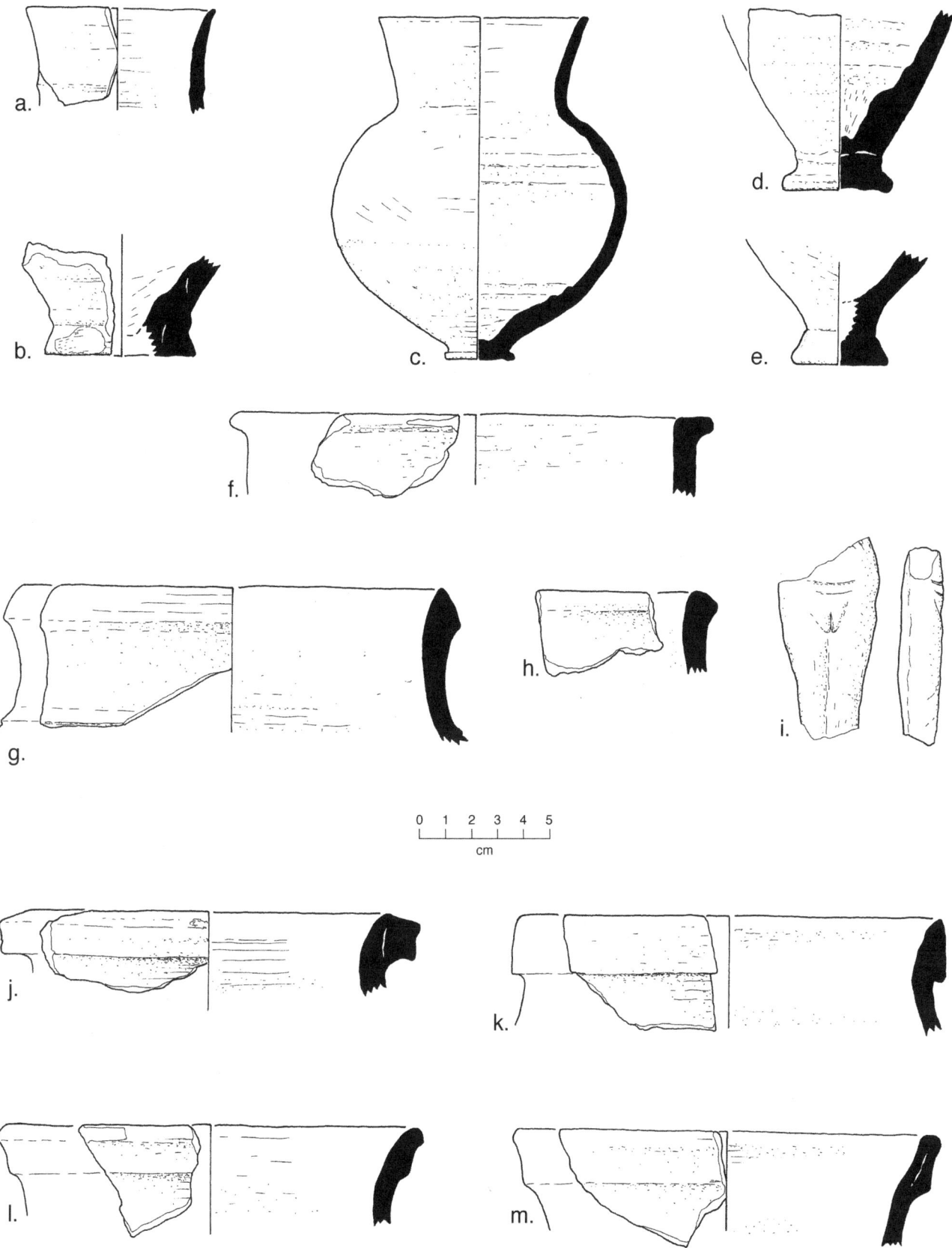

Figure 4.26. Goblet, bowl, and jar sherds from Tepe Patak (DL-35).

Figure 4.27. Basins and large jars from Tepe Patak (DL-35).

a, Basin with band rim (SM-RJ-7) (Area 12). 10% medium sand and vegetal inclusions, Dm 38, RT 1.75, RHt 3.14, ST 0.90, white (2.5Y 8/2) body (Suse Ville Royale AXII [Sukkalmaḫḫu] Group 35b: Pl. 44:4–5).

b, Basin with band rim (SM-RJ-7) (Area 6–7). 10% vegetal and medium sand inclusions, Dm 40, RT 1.38, RHt 3.75, ST 2.12, white (2.5Y 8/1) body (Farukhābād B11–13 [Sukkalmaḫḫu]: Fig. 86L; Suse Ville Royale AXII [Sukkalmaḫḫu] Group 35b: Pl. 44:5).

c, Basin with ledge rim (SM-RB-5) (Area 6–7). 20% vegetal and medium sand inclusions, Dm 36, RT 2.89, ST 1.05, RHt 1.31, greenish-white (5Y 8/2) body (Farukhābād B11–13 [Sukkalmaḫḫu]: Fig. 86m; Suse Ville Royale AXV [Sukkalmaḫḫu] Group 34: Pl. 43:6, BV [Šimaški] Group 35b: Pl. 44:7; Suse Ville Royale I, Level 6 [VA]: Fig. 38:3).

d, Large jar with neckless ledge rim (Area 12). 5% vegetal inclusions, Dm 31, RT 3.08, ST 1.71, RHt 1.81, pink (5YR 7/4) body, very pale brown (10YR 7/3) body (Farukhābād B15 [Šimaški]: Fig. 90h; Suse Ville Royale AXIV–XV [Sukkalmaḫḫu] Group 36: Pl. 45:4, 5, 8, Pl. 46:4).

e, Basin with grooved band rim (SI-RB-4) (Area 12). 5% medium sand and vegetal inclusions, Dm 36, RT 1.20, ST 0.84, RHt 3.18, pale yellow (2.5Y 8/3) body, interior damage (cf. Farukhābād B17–19 [Šimaški]: Fig. 84j).

f, Basin with grooved band rim (SI-RB-4) (Area 6–7). 10% vegetal and medium sand inclusions, Dm 40, RT 1.80, RHt 2.53, ST 0.61, light gray (10YR 7/2) body (Farukhābād B17–19 [Šimaški]: Fig. 84h, j).

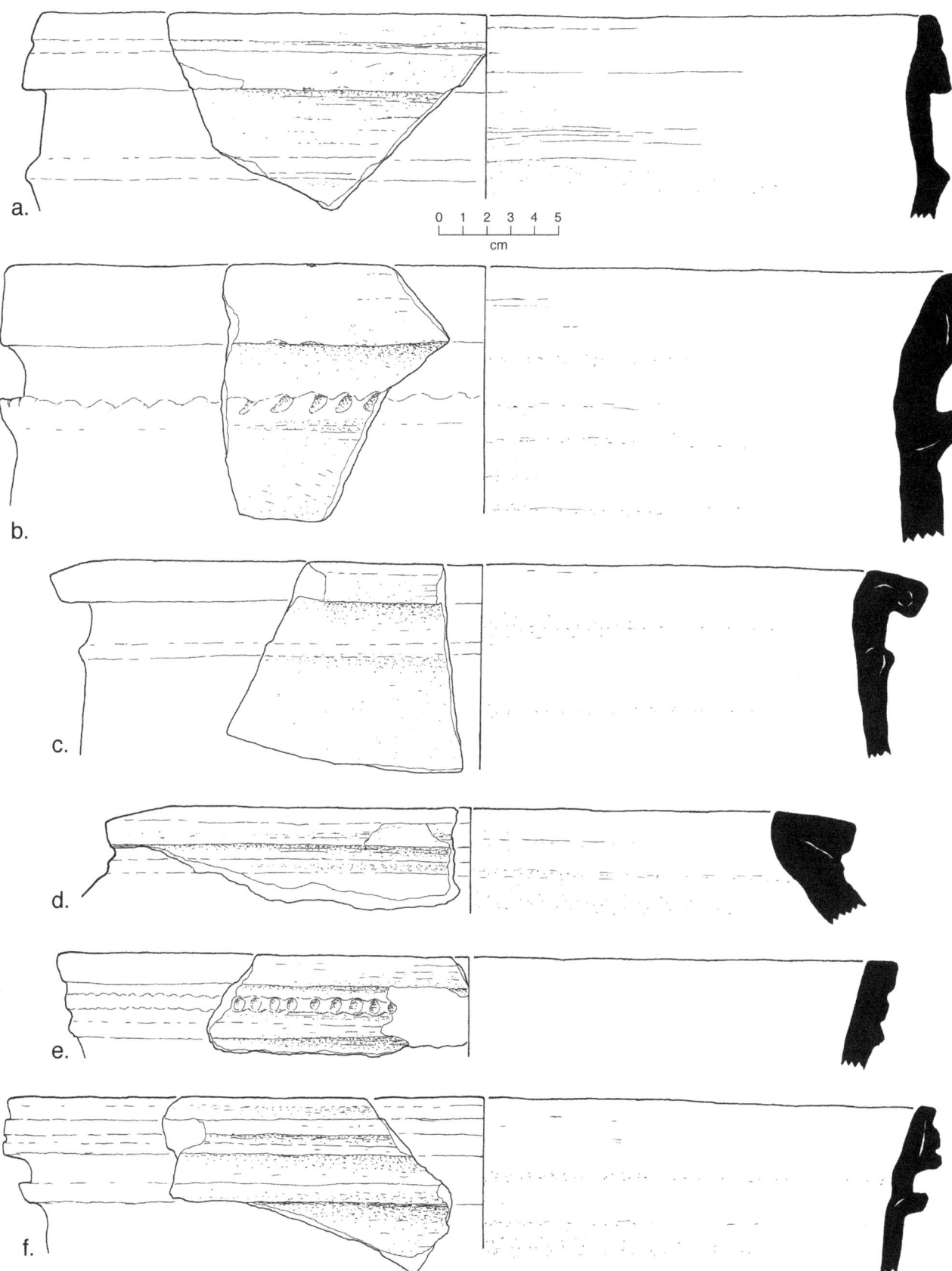

Figure 4.27. Basins and large jars from Tepe Patak (DL-35).

Figure 4.28. Bowls and jars of the mid-Ist millennium B.C. from Tepe Patak (DL-35).

a, Incurved bowl rim (AM-RC-3) (Area 12). 5% fine sand and calcite inclusions, Dm 19, RT 0.56, ST 0.49, pink (7.5YR 7/4) body, very pale brown (10YR 7/3) surface, light interior burnish (Suse Ville Royale II, Level 3D [Seleucid]: Fig. 20:1).

b, Incurved bowl rim (AM-RC-3) (Area 6–7). Trace of calcite and vegetal inclusions, Dm 16, RT 0.44, ST 0.55, red (2.5YR 5/6) body, light exterior burnish (Suse Ville Royale II, Level 3D [Seleucid]: Fig. 20:2).

c, Incurved bowl rim (AM-RC-3) (Area 2). No visible inclusions, Dm 18, RT 0.33, ST 0.27, very pale brown (10YR 7/3) body (Suse Ville Royale II, Level 3D [Seleucid]: Fig. 20:3; Suse Apadana, Level 5f [Late Achaemenid–Early Seleucid]: Fig. 57:4).

d, Carinated bowl with beaded rim (AM-RB-2) (Area 13). 5% fine sand inclusions, Dm 26, RT 1.12, ST 0.87, RHt 1.29, pale brown (10YR 6/3) body, very pale brown (10YR 8/3) surface.

e, Bowl with beaded rim (AM-RB-2) (Area 7). 5% fine sand inclusions, Dm 25, RT 1.40, ST 0.40, RHt 1.25, pale brown (10YR 6/3) body, very pale brown (10YR 8/3) surface.

f, Carinated bowl with beaded rim (AM-RB-2) (Area 13). 15% medium sand inclusions, Dm 25, RT 0.99, ST 0.58, RHt 0.93, pale brown (10YR 6/3) body, very pale brown (10YR 7/3) surface (Suse Ville Royale II, Level 5A [Achaemenid]: Fig. 10:9).

g, Bowl with beaded rim (AM-RB-2) (Area 12). 5% fine sand and vegetal inclusions, Dm ca. 48, RT 1.52, ST 0.58, RHt 2.08, pale brown (10YR 6/3) body (same comparanda as *f*).

h, Carinated bowl with beaded rim (AM-RB-2) (Area 12). 5% fine sand inclusions, Dm 20, RT 1.39, ST 0.43, RHt 1.49, light brown (7.5YR 6/4) body, very pale brown (10YR 8/3) surface, light exterior-interior burnish (same comparanda as *f*).

i, Carinated bowl with beaded rim (AM-RB-2) (Area 15). 10% fine sand inclusions, Dm 34, RT 1.49, ST 0.52, RHt 1.21, reddish-yellow (5YR 7/5) body, white (10YR 8/2) surface (same comparanda as *f*).

j, Carinated bowl with beaded rim (AM-RB-2) (Area 2). 5% fine sand inclusions, Dm ca. 20, RT 1.40, ST 0.52, RHt 1.36, very pale brown (10YR 8/3) surface (same comparanda as *f*).

k, Carinated bowl with beaded rim (AM-RB-2) (Area 14). 15% fine sand inclusions, Dm 38, RT 1.16, ST 0.67, RHt 1.33, pink (7.5YR 7/4) body, very pale brown (10YR 7/4) surface (Suse Ville Royale II, Level 5 [Achaemenid]: Fig. 7:11).

l, Flared neck jar with beaded rim (AM-RJ-3) (Area 14). 10% fine sand inclusions, Dm 10, NT 0.55, RT 1.09, RHt 1.07, pink (7.5YR 7/4) body, very pale brown (10YR 8/3) surface (Suse Ville Royale II, Level 4 [Achaemenid]: Fig. 16:1).

m, Flared neck jar with beaded rim (AM-RJ-3) (Area 17). 10% fine sand inclusions, Dm 11, NT 0.50, RT 1.05, RHt 1.28, pinkish-gray (7.5YR 7/3) body, pale yellow (2.5Y 8/3) surface (cf. Suse Ville Royale II, Level 4 [Achaemenid]: Fig. 16:4).

n, Straight neck jar with beaded rim (AM-RJ-4) (Area 13). 15% fine sand inclusions, Dm 11, NT 0.55, RT 1.09, RHt 1.62, light brown (7.5YR 6/4) body, greenish-white (5Y 8/2) surface (Suse Ville Royale II, Level 4 [Achaemenid]: Fig. 16:5).

o, Flared neck jar with beaded rim and neck ridge (AM-RJ-5) (Area 17). 10% fine sand inclusions, Dm 9, NT 0.50, RT 1.36, RHt 1.45, greenish-white (5Y 8/2) body (Suse Ville Royale II, Level 5B [Achaemenid]: Fig. 16:2, 3).

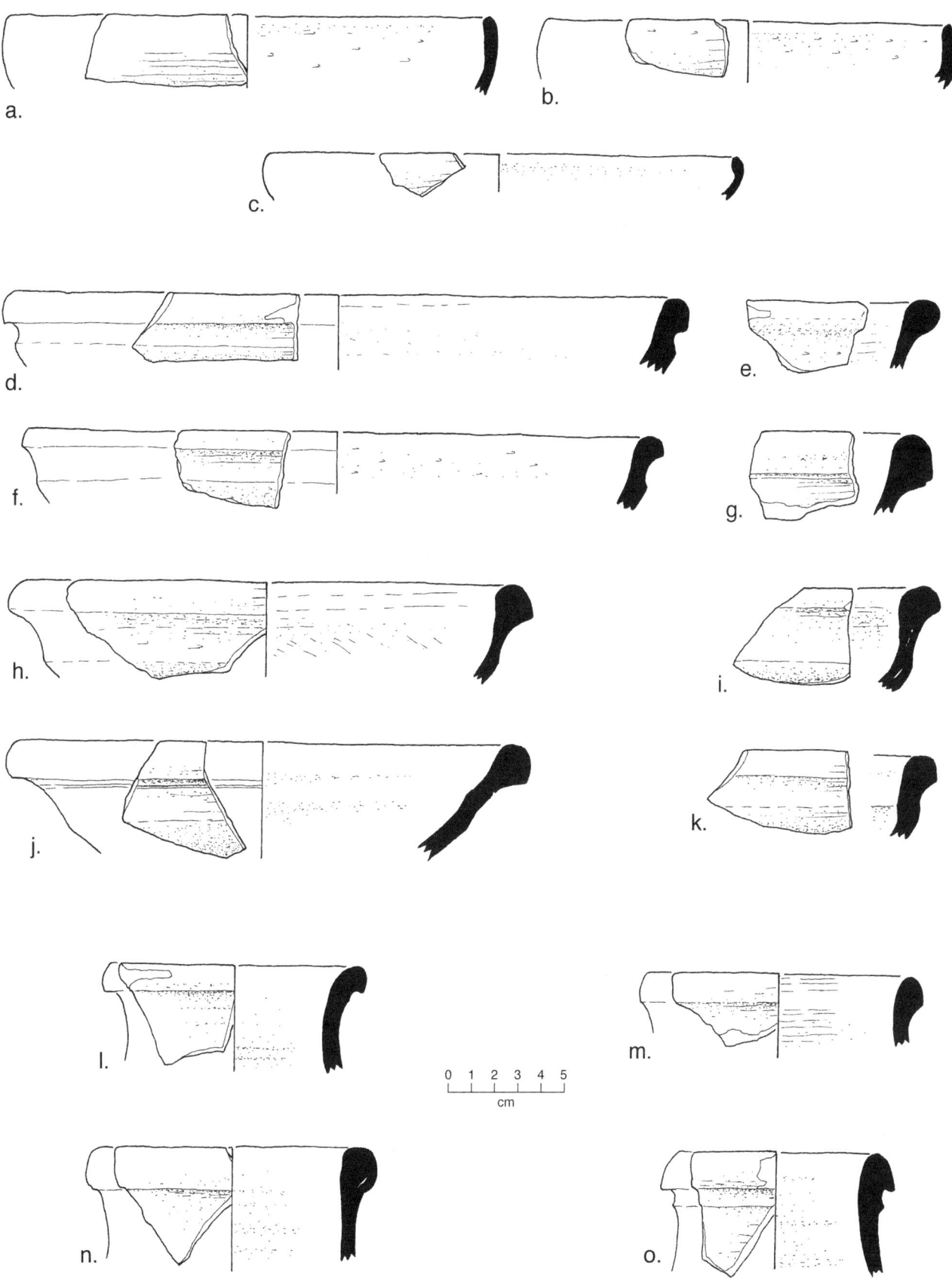

Figure 4.28. Bowls and jars of the mid-Ist millennium B.C. from Tepe Patak (DL-35).

Number: DL-41
Name: DL-41 (Miroschedji DK-25)
Grid: E 47°26'44" / N 32°30'34"
  NIOC: E 1729.1 – N 1170.0
  UTM: E 729, 739 / N 3, 599, 542
Dimensions:
  Length: 570 m (N-S)
  Width: 440 m (E-W)
  Height: 6 m
  Area: 22 ha

Modern Environmental Context: On the Dawairij or Āb-Dānān River alluvial fan, about two km from the present channel of the river.

Features: This site consists of a number of small mounds surrounded by a wall in the form of a rough rectangle with rounded corners (Fig. 4.29, Plate 6).

Phases: The 1968 survey team made only a preliminary visit to this site, and was unable to return to conduct a more careful examination. They recovered two hatched raised strips (ED-O-2) and two round lip jar rims (ED-RJ-1) of Early Dynastic affinity, as well as several Parthian sherds. Fortunately, Miroschedji recovered a diagnostic solid footed conical cup fragment (ED-BC-1: Miroschedji 1981: Fig. 61:3) and rims of small carinated jars (ED-RJ-6) with monochrome decorations (ED-O-6: Miroschedji 1981: Fig. 61:2, 5). The former suggests occupation during the Early Dynastic I–II phase; the other ceramics clearly indicate a Later Early Dynastic III phase occupation. More collections are needed.

Comment: The recognition of the importance of this site is due entirely to Pierre de Miroschedji. As the site is well preserved and is clear on both aerial and satellite imagery, we can provide a new map and better measurements than previously possible. We ascribe the full area, about 22 ha, of this substantial site to the Later Early Dynastic period, and suggest that it was occupied as Musiyān diminished in size late in the Early Dynastic period. The Parthian reoccupation will be discussed in our final volume. See Chapter 6 for a discussion of Canals DL-317C and D.

The Archaeological Sites and Their Interpretation 77

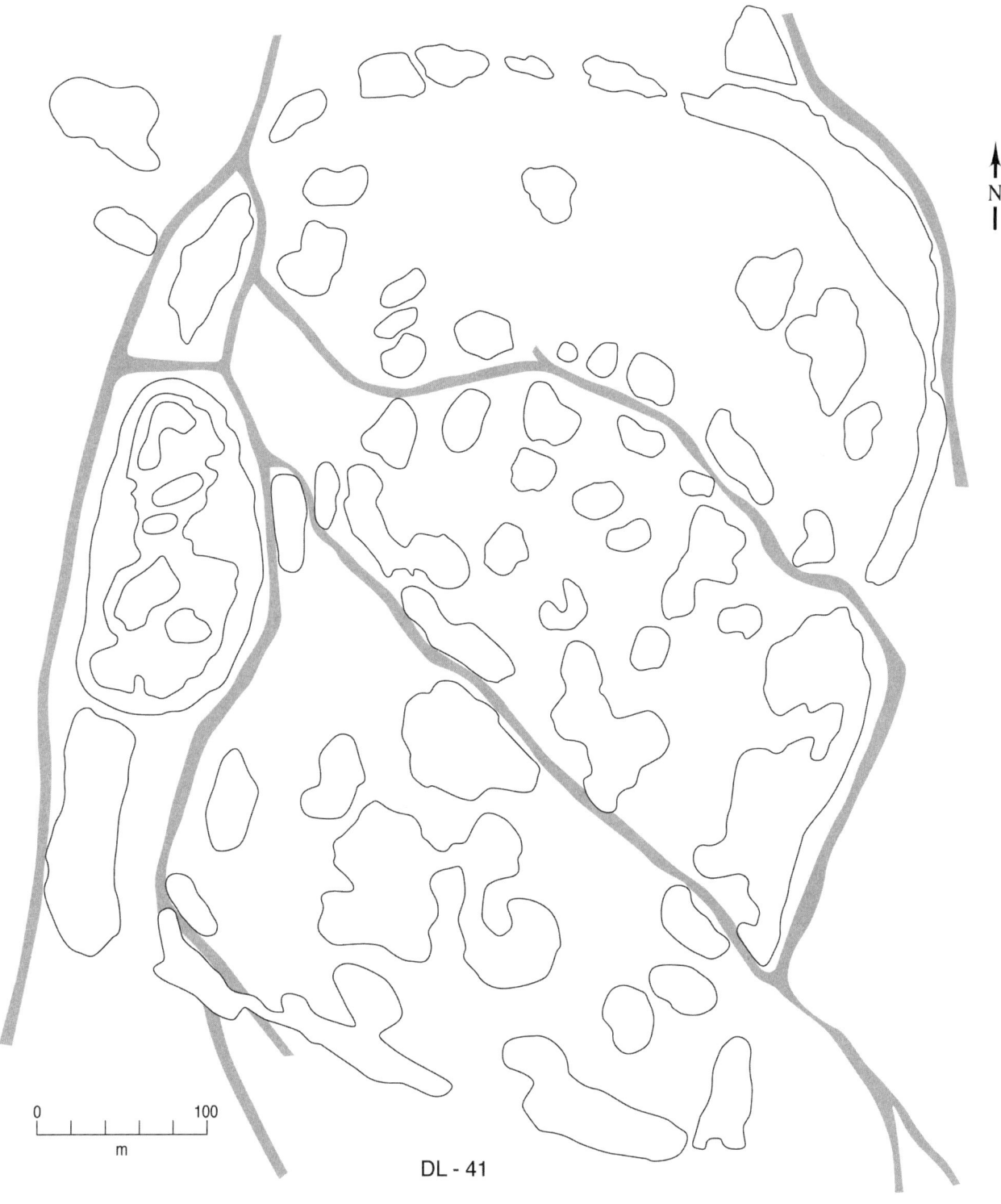

Figure 4.29. Map of DL-41. (Gray lines mark recent canals.)

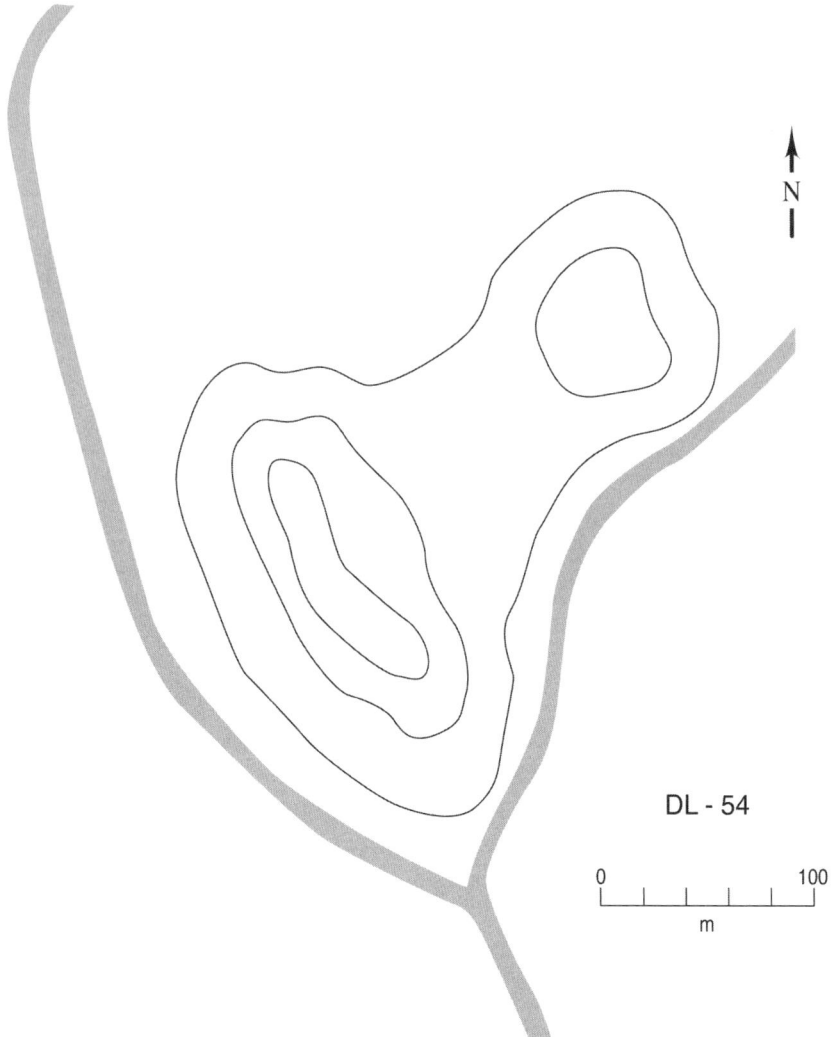

Figure 4.30. Map of Tepe Sohz (DL-54). (Gray lines mark recent watercourses.)

Number: DL-54 (Miroschedji DK-5)
Name: Tepe Sohz (Soza, Suzah)
Grid: E 47°23'50" / N 32°32'53"
  NIOC: E 1726.0 – N 1174.2
  UTM: E 726, 815 / N 3, 601, 787
Dimensions:
  Length: 244 m (NE-SW)
  Width: 175 m (NW-SE)
  Height: 4 m
  Area: 3.2 ha

Modern Environmental Context: On the Dawairij or Āb-Dānān River alluvial fan, about a kilometer and a half northeast of the present channel of the river.

Features: This site consists of a larger southern mound and a smaller northern mound connected by a saddle, but the sherd scatter extended beyond the mound on to the plain (Fig. 4.30).

Phases: The Later Early Dynastic III phase occupation is indicated by the various jars with heavier round lip (ED-RJ-2: Neely and Wright 1994: Fig. IV.33*i–k*) and band rim (ED-RJ-5: Neely and Wright 1994: Fig. IV.33*l, m*) with hatched strip decoration (ED-O-2: Neely and Wright 1994: Fig. IV.33*p*). Carinated bowls (ED-RB-4: Neely and Wright 1994: Fig. IV.33*q, r*) and small carinated jars (ED-RJ-6: Neely and Wright 1994: Fig. IV.33*s*) with monochrome designs (ED-O-3: Neely and Wright 1994: Fig. IV.33*s–u*) were found. These ceramics covered the surfaces of the two mounds and extended outward a few meters onto the alluvial plain.

Comment: A small occupation of the Earlier Early Dynastic I–II phase is arbitrarily assigned 1.0 ha of area, which is covered by later deposits. We ascribe the full 3.2-ha area of this site to the Later Early Dynastic III phase. See Chapter 6 for a discussion of Canal 317C.

References: Neely and Wright 1994:105–9; Miroschedji 1981b:173, 175.

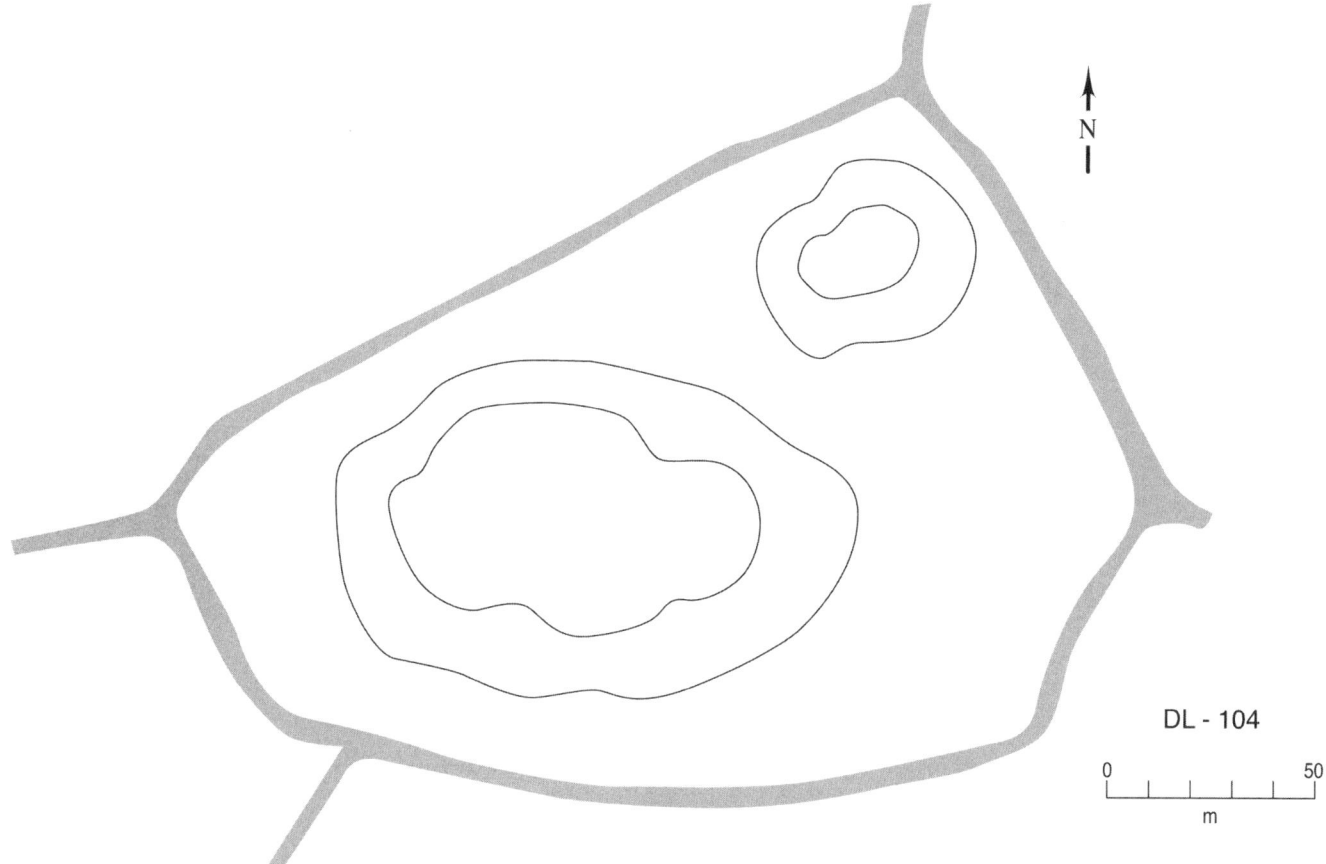

Figure 4.31. Map of Tepe Mohr (DL-104). (Gray lines mark recent canals.)

Number: DL-104
Name: Tepe Mohr
Grid: E 47°19'02" / N 32°32'07"
    NIOC: E 1717.1 – N 1171.8
    UTM: E 717, 644 / N 3, 602, 138
Dimensions: (West mound only)
    Length: 120 m (NW-SE)
    Width: 75 m (NE-SW)
    Height: 3.25 m
    Area: 0.6 ha

Modern Environmental Context: At the junction of the southern alluvial slopes of the plain and the alluvial fan of the Dawairij or Āb-Dānān River, about seven km southwest of the present Dawairij channel and six km east of present Mehmeh channel.

Features: This site consists of two mounds, but late IIIrd millennium material was restricted to the southwest part (Fig. 4.31). No evidence of architecture was visible on the surface of the site. Recent satellite images show that the site has been bulldozed, and that trimming the west mound on the north and south has left a rectangular remnant, and cutting away the east half of the east mound has left a standing section facing east.

Phases: The identification of Later Early Dynastic III occupation rests on heavier examples of the round lip jar (ED-RJ-1: Neely and Wright 1981: Fig. IV.42L) and ledge rim jar (ED-RJ-4: Neely and Wright 1981: Fig. IV.42o), rather than on unequivocally diagnostic types such as the small carinated jars (ED-RJ-6) with monochrome decoration (ED-O-3).

Indicative of the Šimaški phase are an indented rim carinated bowl or cup (SI-RC-3: Fig. 4.32a), two larger ribbed basins with flat lips (SI-RB-5: Fig. 4.32c, d), and a band rim jar (Fig. 4.32b), reported before as Early Dynastic, but which may be

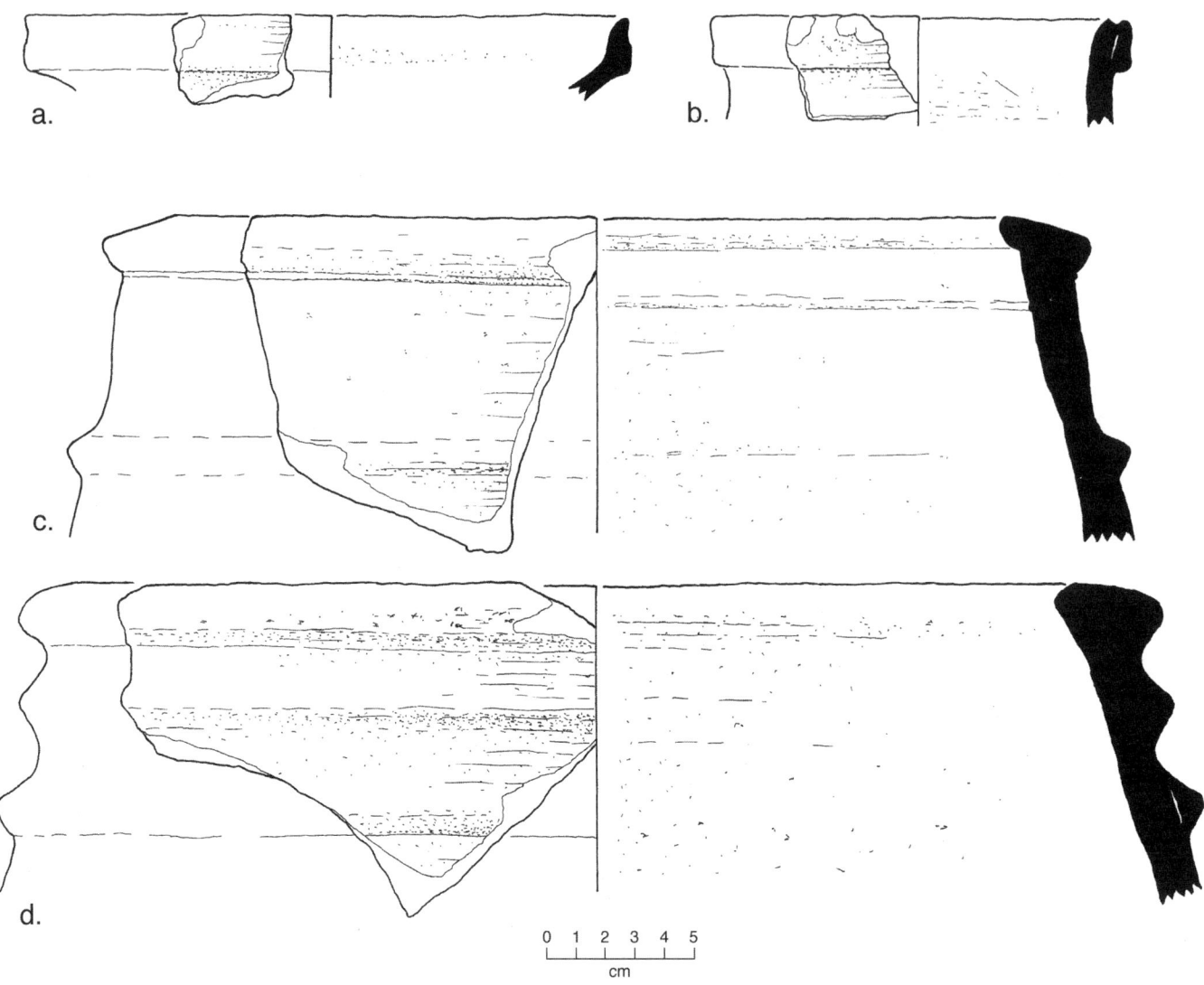

Figure 4.32. Late IIIrd millennium B.C. bowl, jar, and basins from Tepe Mohr (DL-104).

a, Indented rim bowl (SI-RC-3) (Area 1). 15% medium sand inclusions, Dm 18, ST 0.68, RHt 1.72, light reddish-brown (2.5YR 6/4) body, very pale brown (10YR 7/3) surface (Farukhābād B16–17 [Šimaški]: Fig. 84b; Suse Ville Royale BVII–VI [Šimaški] Group 7a: Pl. 7:3, 10; Suse Ville Royale I, Level 6 [VA]: Fig. 38:3, Level 3 [VB]: Figs. 45:6, 49:10, 11).

b, Band rim jar with lip groove (Area 1). 15% medium sand inclusions, Dm 12, RT 1.24, ST 0.80, RHt 1.55, white (2.5Y 8/2) body.

c, Basin with flat lip (SI-RB-5) (Area 1). 10% medium sand and vegetal inclusions, Dm 30, RT 2.42, ST 1.25, RHt 1.84, very pale brown (10YR 8/3) body (Suse Ville Royale I, Level 3, 4A [VB]: Fig. 51:5; cf. Suse Ville Royale II, Level 10 [Later Middle Elamite]: Fig. 15:19).

d, Basin with flat lip (SI-RB-5) (Area 2). 10% fine sand, calcite, and vegetal inclusions, Dm 34, RT 3.14, ST 1.12, pink (5YR 7/4) body (Farukhābād B15 [Šimaški]: Fig. 90h, cf. B6 [Early Middle Elamite]: Fig. 89q; Suse Ville Royale I, Level 3 [VB]: Fig. 51:5).

Šimaški. The small number of items suggests a brief or limited occupation.

While visiting later site DL-103, a short distance northeast of DL-104, a shepherd gave Neely the head of the demon *Pazuzu* (Plate 8 *lower*) of gray calcite. The round nose, feline whiskers, short beard, and large ears are similar to other representations from Mesopotamia dating to the first millennium B.C. (Heessel 2001). The unique drilled eyes and grooved mouth of the Deh Lurān piece may have contained insets. When asked where he found this, the shepherd gestured across the whole central Deh Lurān Plain. While Gārān (DL-34), the largest site of this time, is a likely find spot, the figurine may have been lost and found many times, and its most recent specific find spot may not have been important. The only other *Puzuzu* head from Iran was excavated at Nush-i Jan (Heessel 2001:38), although there are other unprovenienced examples in the Tehran museum (Muscarella 1985:729). The four lines of damaged cuneiform signs on the back of the head are not fully readable; most probably this is another example of pseudo-epigraphic incisions meant to look like cuneiform signs, analogous to Heessel nos. 1, 107, and 113.

Comment: We ascribe an area of 2.0 ha to this small site to both the Earlier Early Dynastic I–II phase and the Later Early Dynastic III phase. For the limited Šimaški phase occupation, we suggest an area of about 0.3 ha. See Chapter 6 for a discussion of Canal DL-115B.

References: Gautier and Lampre 1905:81, Fig. 9; Neely and Wright 1994:124–30; Miroschedji 1981:173, 175.

Chapter 5

Early Historic Settlement Patterns on the Deh Lurān Plain

Henry T. Wright

The excavations at Tepe Farukhābād and Susa have resulted in the definition of a sequence of seven successive cultural phases dating between the mid-IIIrd and mid-Ist millennium B.C. These excavations provide us with some evidence of changing economic and social life. However, complementary evidence of changes in regional population and organization can come only from archaeological survey. In this chapter we consider the archaeological evidence as if they were prehistoric cultures, lacking epigraphic evidence. In Chapter 7, this archaeological evidence will be contrasted and integrated with the textual evidence.

The Later Early Dynastic Phase

The period from 2550 to 2300 B.C. was one of intense competition between regional states, leading to the emergence of the first experiments with the building of transregional polities. During this time span, the stylistic relations with Pusht-i Kuh in general and the Diyala area to the northwest (Emberling 1997) gave way to a closer relation with Susiana and southern Iran to the east (Carter and Stolper 1984). Ceramics of this phase were found on eight sites, currently estimated to cover a total of about 46.0 ha (Table 5.1; Fig. 5.1). If this is so, and given the population density suggested by the architecture of the Early Dynastic I–II phase, the population of the Deh Lurān Plain during the Early Dynastic III phase would have been about 9200 people. However, the growth of a short-lived mid-IIIrd millennium center (DL-41) toward the southeast end of the plain—just about the time that Farukhābād (DL-32) and other sites were abandoned, and Musiyān may have become smaller—suggests population displacement within the plain. If so, perhaps about 25 ha of settlement would have been occupied at one time, and the population of the plain would have been about 5000.

Settlement has shifted to the southeast. On the formerly prosperous northwestern part of the plain, the only larger settlement was Baulah (DL-24). Surrounding cultivated fields were watered by the sulfurous but reliable Ab-i Garm, as the Deh Lurānis do today. Farukhābād (DL-32), reduced in size on its high mound, must have drawn water from the brackish Mehmeh River, which had not yet cut down below plain level (Kirkby 1977). All other settlements took water from the sweet waters of the Dawairij River, more recently termed the Āb-Dānān.

Musiyān (DL-20), the established center of the plain since the beginning of the IIIrd millennium, and the small village of Mohr (DL-104) shared a long canal taking off from the right bank, first initiated during the Khazineh phase during the late VIth millennium B.C. (Neely and Wright 1994:107, 190–91). Nearby Gārān (DL-34) may also have used Dawairij or Āb-Dānān water, but it was small enough that it might have used wells and winter runoff from the gravel slopes of the plain. The large village of Sohz (DL-54) as well as the newly emergent large center of DL-41 and the small site of Patak (DL-35) were perhaps gaining population from Musiyān and settlements farther to the northwest. They utilized a long canal taking off from the left bank, one initiated early in the IIIrd millennium B.C. (Neely and Wright 1994:180, 192–93).

Other than their size and location, we know little of most of these Later Early Dynastic settlements. The layers of this period were eroded at Farukhābād (DL-32), but show a contrast of modest and elaborate buildings made with plano-convex mud bricks. Limited excavation at Farukhābād shows that the agricultural economy continued to be focused on goats and sheep, with an added emphasis on probably domesticated pigs. There

83

Table 5.1. Later Early Dynastic settlements.

Site No.	Name	Area (ha)
Centers		
20	Musiyān	ca. 13.0
24	Baulah	4.2
41	–	ca. 20.0
Villages		
32	Farukhābād	ca. 2.5
34	Gārān	ca. 1.5
35	Patak	–
54	Sohz	ca. 2.3
104	Mohr	ca. 2.0

Table 5.2. Šimaški settlements.

Site No.	Name	Area (ha)
Centers		
20	Musiyān	ca. 15.0
24	Baulah	ca. 4.2
Villages and Hamlets		
27	Tenel Ramon	ca. 1.0
32	Farukhābād	ca. 0.5
34	Gārān	ca. 1.8
35	Patak	ca. 1.0
104	Mohr	ca. 0.3

are no plant remains from this period. Major chert processing also occurred at Farukhābād (in spite of its distance from the sources to the east of Deh Lurān) as well as bitumen processing and copper smelting. Even though Farukhābād was declining in importance, its inhabitants received such exotic goods as lapis lazuli from Afghanistan. Surface evidence indicated that people at Sohz (DL-54) concentrated on chert processing, not surprising for a village close to the sources of chert east of the Deh Lurān Plain. Unfortunately, even less is known of the architecture and special activities of the larger centers of Musiyān (DL-20) and DL-41. At the former, Gautier and Lampre (1905) report a large building of the plano-convex brick typical of this time in the center of the site, but provide no plan.

In sum, the archaeological evidence indicates that the prosperous local polity of the Earlier Early Dynastic period persisted, but with some abandonment of smaller villages to the west and a shift of the major center eastward late in the period. These changes support the stylistic evidence of closer relations with Susiana, and suggest that external threats to Deh Lurān came from the west during the Later Early Dynastic phase.

The Šimaški Phase

During the span of time from 2200 to 1900 B.C., the people of Deh Lurān used material items with stylistic relation with both Susiana and Lower Mesopotamia proper. Ceramics of this phase, well known both locally from Farukhābād and from the distant centers of Susa and Nippur, occur on seven sites covering 23.8 ha (Table 5.2; Fig. 5.2). The architecture found in the small hamlet of Farukhābād was modest, but it would be dangerous to extend assessments of this architecture to other sites, particularly to the large and prosperous center at Musiyān. However, the area of excavated Šimaški housing at Susa in the Ville Royal B excavation (Gasche 1973: Plan 6), also modest, gives a useful indication of density. In the central portion of Level BVII, excavation of 485 m^2 revealed major portions of three multi-room units with 145 m^2 of roofed space, enough to house about 15 people, indicating a population of 310 people per hectare. In the central portion of Level BVI, excavation of 305 m^2 revealed major portions of two multi-room units with 94 m^2 of roofed space, enough to house 10 people, indicating a population of 330 people per hectare. Assuming public buildings and other non-domestic structures occupied one-third of the space in these centers, we estimate a population density of 210 persons per hectare, or 4030 persons for the 19.3 ha of the centers. Assuming a population of 115 people per hectare for the smaller villages, following our estimate for earlier villages and hamlets (Neely and Wright 1994:33, 173), we estimate about 485 people in the small settlements. We can therefore suggest that the settled population of the plain was about 4500 people, slightly smaller than that estimated for the preceding Later Early Dynastic phase. This is an average estimate that does not take into account fluctuations due to periods of prosperous growth and immigration versus periods of decline, plague, conflict, and emigration that may have afflicted the plain during the three centuries of the Šimaški span.

During this period, Musiyān was once again the major center of the plain, and the two nearby smaller settlements of Gārān (DL-

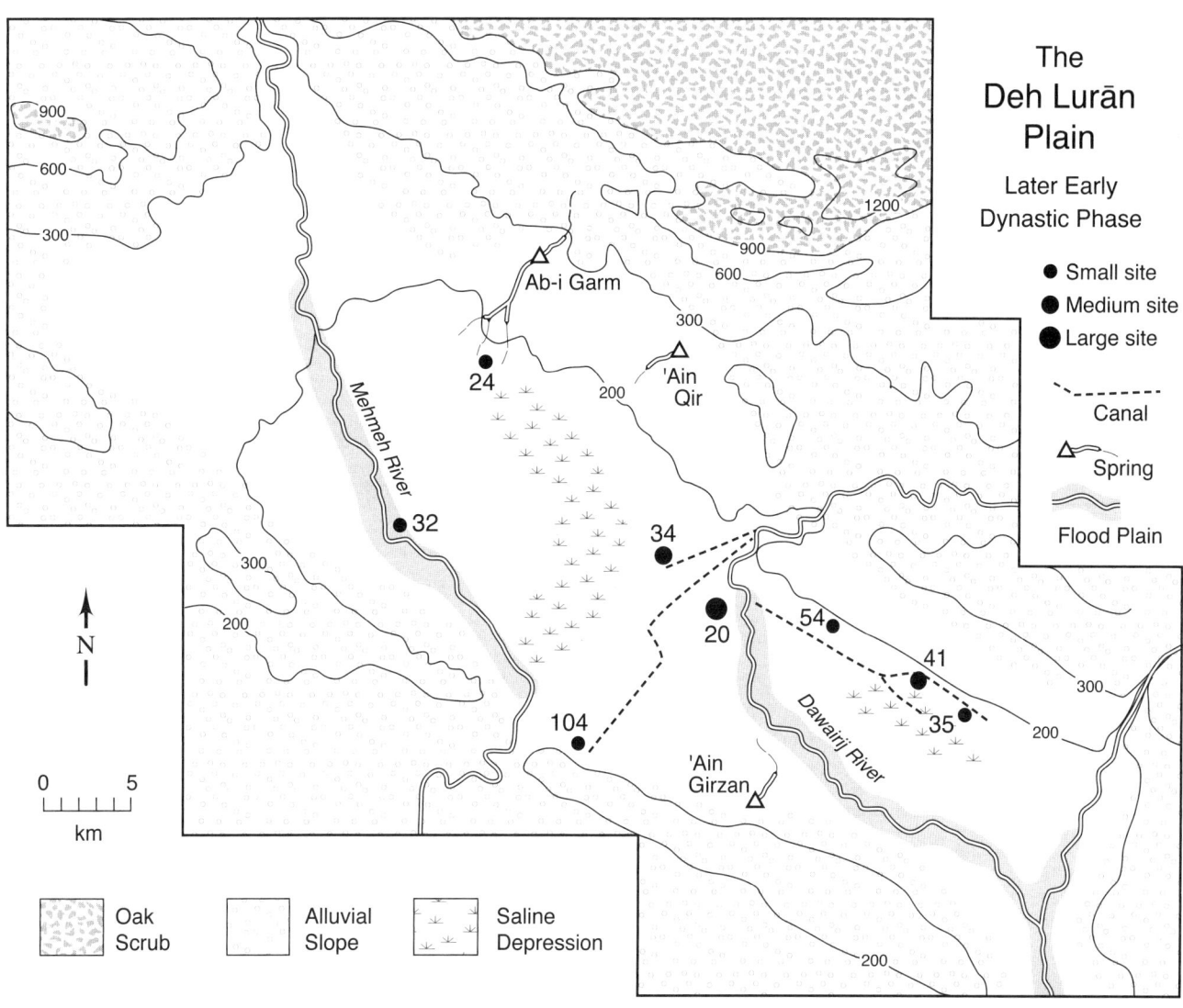

Figure 5.1. Settlements of the Later Early Dynastic phase.

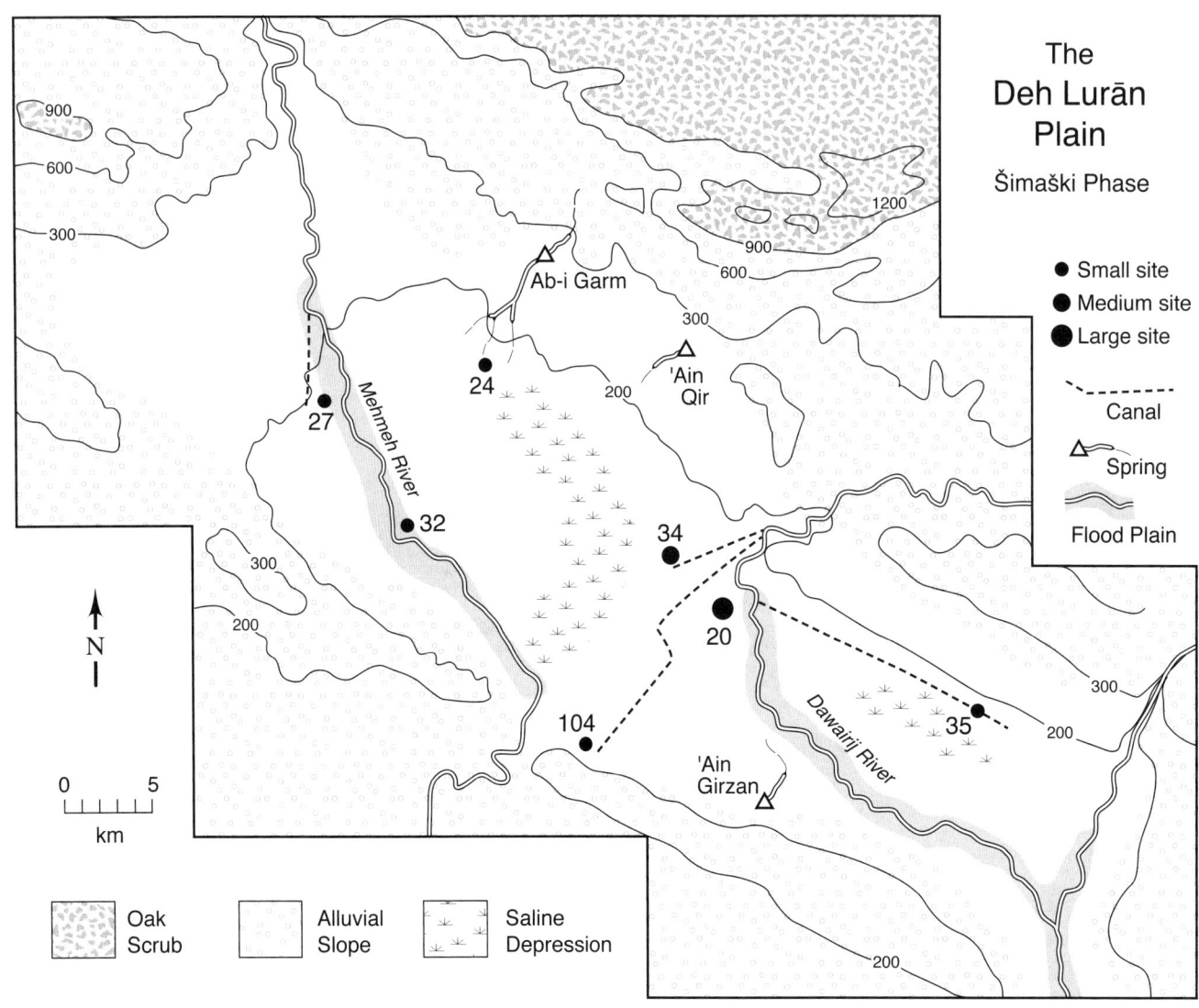

Figure 5.2. Settlements of the Šimaški phase.

34) and Mohr (DL-104) continued to be occupied. Settlements to the southeast taking water from the left bank of the Dawairij or Āb-Dānān, however, were abandoned. To the northwest, the small center of Baulah (DL-24) continued and the settlement at Farukhābād shrank to hamlet size. However, Tenel Ramon (DL-27), west of the Mehmeh, was reoccupied for the first time in centuries. Patak (DL-35) also had limited occupation.

The evidence from excavations is limited. At the hamlet of Farukhābād, only Excavation B contained evidence of Šimaški phase architecture, a sequence of modest single-story buildings with adjacent courtyards. While the domestic pottery assemblage has broad parallels in nearby regions, the rest of the technology was limited. There were a few grinding and chopping tools. Most small cutting tools were expedient flakes of a dark brown chert obtainable from the Mehmeh River, a few tens of meters from the settlement. There is little evidence of the processing of medium gray chert from the area to the southeast of the Deh Lurān Plain, and little evidence of bitumen processing. Even though there were only a few bits of cuprous metal, the scarcity of chert sickle blades suggests that metal sickles were used, but were carefully recycled. The animal bones indicate that sheep, goats, and some cows were kept, but there was little hunting. Thus, Farukhābād was a basic farming community with little evidence of craft, exchange, or social complexity. In contrast, the main center of Musiyān was probably walled and had an elaborate central public building. Unfortunately, we know nothing about domestic organization, crafts, exchange, or politics at this major settlement.

In sum, settlement was similar in extent and size variation to that of the preceding Early Dynastic period. The primary difference is a lack of settlement in the southeast and a shift in observed settlement to the northwest. This suggests that, during this span of time, the external threats to Deh Lurān came from the east. Given the limited evidence we have from survey and from a small excavation in a minor village, there is little direct evidence of control by the Akkadian, Ur III, and Šimaški polities.

The Sukkalmaḫḫu Phase

The stylistic evidence from Deh Lurān indicates that from 1900 to 1600 B.C., Deh Lurān was closely related to Susa, a capital of the powerful Sukkalmaḫḫu Elamite polity. Ceramics of this phase, well known both locally from Farukhābād and from Susa to the east, occur on six sites covering 28.2 ha (Table 5.3; Fig. 5.3). The limited exposure of architecture found in the hamlet-sized settlement of Farukhābād is not helpful for assessing population density. However, the area of excavated Sukkalmaḫḫu housing at Susa in the Ville Royal A excavation (Gasche 1973: Plans 4 and 5) gives a useful indication. In the west sector of Level AXV, excavation of 3500 m² exposed major portions of at least 29 multi-room domestic units (some large and some small and fitted into irregular spaces), one shrine, and some streets. In the probable domestic units, there are about 1150 m² of roofed

Table 5.3. Sukkalmaḫḫu settlements.

Site No.	Name	Area (ha)
Centers		
20	Musiyān	15.0
24	Baulah	4.2
35	Patak	ca. 5.0
Villages and Hamlets		
27	Tenel Ramon	2.0
32	Farukhābād	ca. 0.2
34	Gārān	ca. 1.8

space, enough to house 115 people, indicating a population of 328 people per hectare. In Level AXIV, excavation of 5100 m² revealed major portions of 6 substantial multi-room units varying in size from 100 to at least 570 m². In the area excavated to Level AXIV, there is a total of 1180 m² of roofed space, enough to house about 120 people. Many of the walls are a meter or more thick, strong enough to bear another story, but there are few indications of stairs to an upper story. If we allow for upper story rooms adding another third to the roofed space, the excavated area could have housed 160 people, indicating a population of about 245 to 320 people per hectare. As we have already accounted for non-domestic structures and streets in these relatively large excavations, we estimate 300 people per hectare for the Sukkalmaḫḫu phase as a rough average population density, or 7800 for the 26.2 ha comprising the major center of Musiyān (DL-20) and the possibly walled smaller centers of Patak (DL-35) and Baulah (DL-24), as well as the village of Tenel Ramon (DL-27). Assuming 115 persons per hectare for the smaller settlements, this yields an estimate of about 460 people. We can therefore suggest that the maximum settled Sukkalmaḫḫu population of the plain was about 8260 people, almost double that estimated for the preceding Šimaški phase. This estimate depends on the assumption that architectural patterns and population densities of Sukkalmaḫḫu Deh Lurān were similar to those of contemporary Susa. In addition, the static technique utilized gives an estimate assuming that all settlements were occupied at the same time.

During this period, Musiyān continued to be the major center of the plain, probably taking its water from the right bank of the Dawairij or Āb-Dānān. The nearby small settlement of Gārān (DL-34) could have taken water from the 'Ain Qir spring or the Dawairij. There were two subsidiary centers, both with evidence

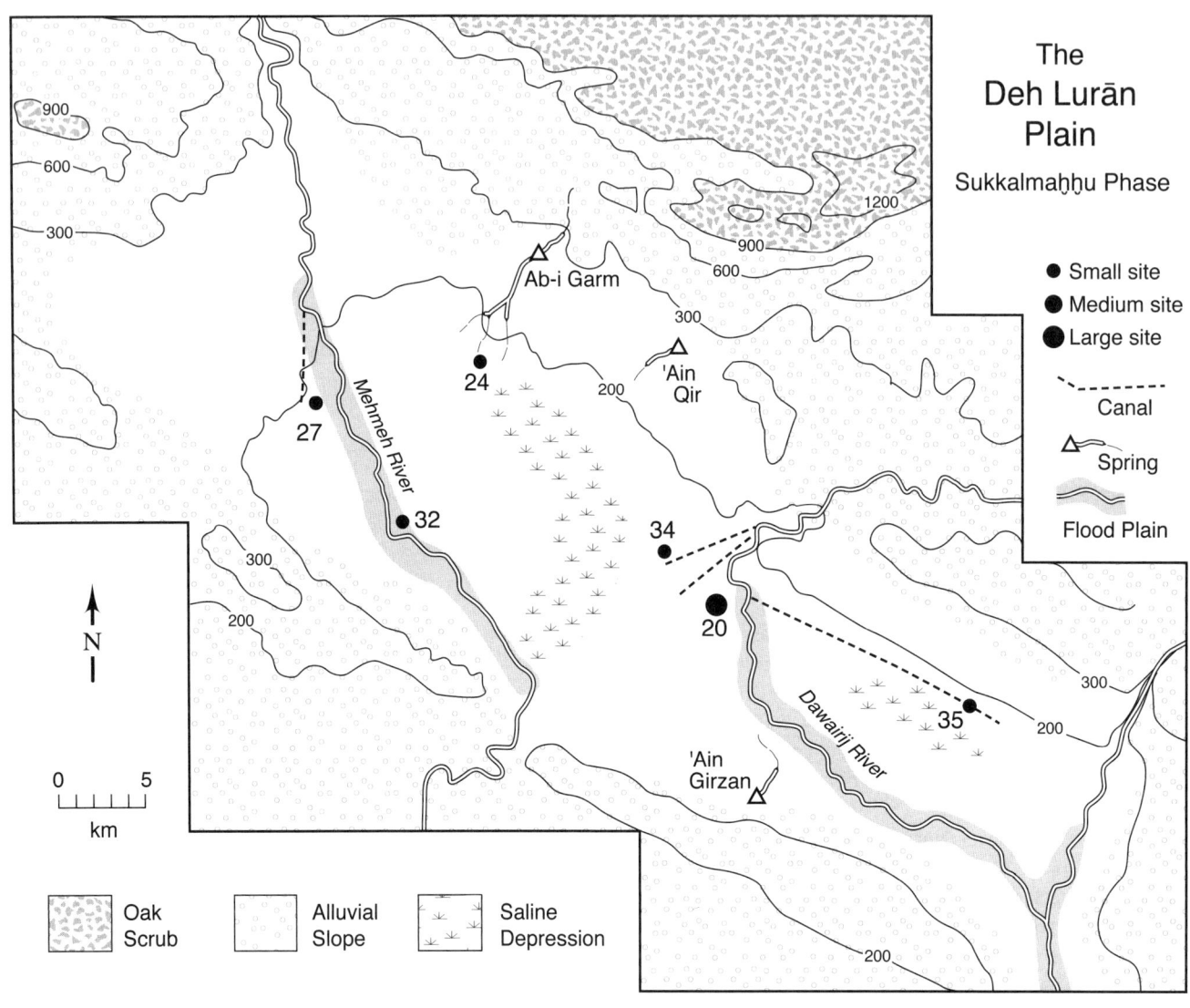

Figure 5.3. Settlements of the Sukkalmaḫḫu phase.

of fortification, perhaps of Sukkalmaḫḫu date. Baulah (DL-24), about 21 km to the northwest, could have relied on the Ab-i Garm for its water. Patak (DL-35), about 14 km southeast of Musiyān, could have gotten water only by a long canal taking off from the left bank of the Dawairij. The settlement high on the mound of Farukhābād (DL-32) was very small but was fortified, probably to guard the Mehmeh River crossing, as perhaps was Tenel Ramon (DL-27), just west of the Mehmeh.

At the small settlement of Farukhābād, Excavation B contained provocative evidence of the Sukkalmaḫḫu phase. The top of the mound had a massive mud brick revetment surrounding an area more than 55 by 30 m. Unfortunately, the evidence of any buildings within this small fort remains buried beneath the remains of the subsequent Early Middle Elamite hamlet and a much later Parthian structure (Gautier and Lampre 1905:84; Wright ed. 1981:224). However, a larger pit just outside the revetment produced a flexed burial and a range of artifacts. As in Šimaški times, most small cutting tools were expedient flakes of a dark brown chert obtainable from the Mehmeh River, a few tens of meters from the settlement. There is no evidence of any craft activities. Sheep, goats, and surprising numbers of cows were consumed, but there was little hunting. A significant find from this small fortified settlement are series of bones of a domestic horse, with evidence of a lip ring rather than a bit (Redding 1981:243–44, Pl. 21), and thus probably used as a pack or courier animal. There is no direct evidence of plant foods, but the lack of both sickle blades and grinding stones suggests little grain production or processing. It is possible that the diminution of the site and the lessened evidence of cultivation is because the lower Mehmeh had entrenched into the alluvial plain (cf. the geological discussion in Kirkby 1977, where an earlier date is proposed) and was no longer easily used for small-scale irrigation. Alternately, perhaps Farukhābād was an outpost overlooking the southern and western approaches to the Deh Lurān Plain, sustained with rations of animals and other products from the main center of Musiyān. It is unfortunate that we know little about the walled center of Musiyān and its elaborate central public building during this time.

Thus, the Sukkalmaḫḫu phase was a time of relative prosperity and high population on the plain. Settlements were widespread and relatively large, and a majority appear to have been fortified. We can infer that the Sukkalmaḫs were attempting to maintain a strong presence on their western border, probably to guard against incursions from the rulers of the Old Babylonian polities to the west.

The Earlier Middle Elamite Phase

From 1600 B.C. until about 1300 B.C., the stylistic evidence of the ceramics indicates relations with both Earlier Middle Elamite Susiana to the east and later Old Babylonian Mesopotamia to the west. Ceramics of this phase, well known both locally from Farukhābād and from the great centers of Susa and Haft Tepe, ancient *Gapnak*, on the Susiana Plain to the east, occur on five

Table 5.4. Earlier Middle Elamite settlements.

Site No.	Name	Area (ha)
Centers		
20	Musiyān	ca. 10.0
34	Gārān	ca. 12.0
35	Patak	ca. 5.0
Villages and Hamlets		
27	Tenel Ramon	ca. 2.0
32	Farukhābād	ca. 0.2

sites covering about 29 ha (Table 5.4; Fig. 5.4). However, since Musiyān was diminishing and Gārān was growing during this period, it is likely that the actual settlement area was lower, about 19 ha. The tiny exposure of architecture found in the hamlet-sized settlement of Farukhābād is, as for the preceding phase, not helpful for assessing population. In addition, the areas of excavated Early Middle Elamite architecture at Susa in the Ville Royal A excavation (Gasche 1973: Plans 2, 3, and 4) may not be occupied simultaneously. Thus, we hesitate to infer population density from the roofed area for the previous phase. Instead, using the figure of 300 people per hectare discussed above, we can suggest a maximum population of about 5700 people.

During this phase, Musiyān diminished, perhaps signaling that the down-cutting of the Dawairij or Āb-Dānān had reached the upper plain and reduced Musiyān's water supply. A newly emergent center developed at the established village site of Gārān. Its remains are deeply buried under Achaemenid and Hellenistic remains, and without excavation we cannot tell whether it was fortified. With a larger population, perhaps 3000 or more, however, it would probably have exceeded the limited water supplies available from wells, from the harvesting of runoff, or from long canals from the small springs of Ab-i Garm—no longer needed for now abandoned Baulah (DL-24)—or of 'Ain Qir. If, however, the copious waters of the Dawairij were used, given that it has by this time downcut below plain level, a canal would have been difficult without the use of the *qanāt*, a vented tunnel. The arguments for the first appearance of *qanāts* will be discussed in Chapter 6. The smaller center of Patak appears to have been fortified during this period. Patak would have continued to get water from a canal taking off from the left bank of the Dawairij, which may also have received a *qanāt* offtake. To the west, the hamlet at Farukhābād (DL-32) remained tiny. But the village at Tenel Ramon (DL-27) remained substantial, and was perhaps fortified.

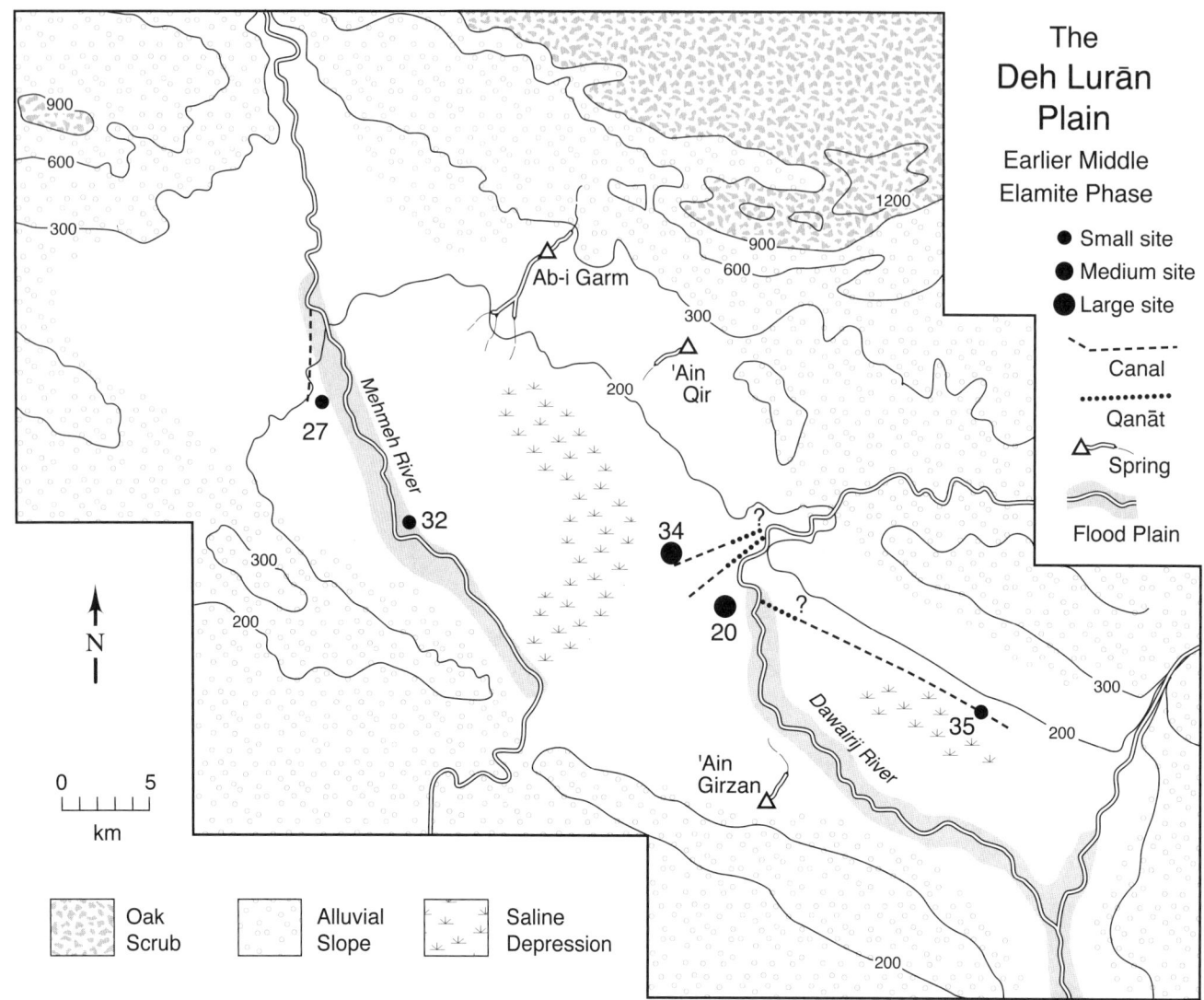

Figure 5.4. Settlements of the Earlier Middle Elamite phase.

Our only excavated samples are from Farukhābād, which at this time was a tiny hamlet perched on the ruins of the Sukkalmaḫḫu fort. Wall stubs and floors in the northeast section of Excavation B (Wright ed. 1981: Fig. 83) suggest a few closely spaced domestic structures with well-laid mud brick walls. There is an oven with ashy fill, and the shaft of a well or drain. The upper fill of the latter contains pottery vessels and fragmentary human bones. The faunal remains are similar to those of the Šimaški phase, and wheat and barley are attested. There is no evidence of craft or exchange. There are some moldmade female figurines identical to those of sites on the Susiana Plain. At least in the area of domestic ceramics and household ritual, the people of this small community seem to have been close to the Elamite world (Carter 1981:216–18, Pl. 20). Though Farukhābād still maintained its commanding view of the western plain, it has once again become a village community of the most basic sort. Unfortunately, we know little about the larger settlements other than what can be inferred from survey.

Clearly Deh Lurān is less populous and perhaps less prosperous than in Sukkalmaḫḫu times. It is possible that increasing aridity during the IInd millennium has led to down-cutting, which made irrigation more difficult. Clearly the exact timing of these changes and of the development of *qanāts* is a subject requiring geoarchaeological research. Whatever the environmental situation, even with the shift of the main center from Musiyān to Gārān, regional organization (Fig. 5.4) was little different from that during the Sukkalmaḫḫu phase.

The Later Middle Elamite Phase

The Later Middle Elamite rulers left a rich record around the traditional capital at Susa as well as the newly built capital and religious center at Choghā Zanbil. Their ceramics are well known, as are the ceramics of contemporary Cassite settlements in Lower Mesopotamia. Only one site, Gārān (DL-34), has produced a few sherds that may be of this period. Their scarcity on a site that has copious remains of both earlier and later periods suggests that Later Middle Elamite or Cassite occupation was ephemeral or short-lived at best. It is possible that outposts were maintained here, but for much of the later IInd millennium, the Deh Lurān Plain must have been a buffer zone occupied only by nomads.

The Neo-Elamite Phase

The early Ist millennium B.C. was a period during which the earlier Iron Age empires fought in the northern part of Lower Mesopotamia, historical Babylonia, and in the Zagros front ranges, as had their Bronze Age predecessors. To what extent Neo-Elamite rulers—centered at Susa until its destruction by the Assyrians in 647 B.C.—reoccupied the Deh Lurān Plain as their western frontier is a question for archaeologists; Miroschedji (1981b) resurveyed the area specifically to assess this point. Neo-Elamite ceramics are known from Miroschedji's (1981b) Ville Royal II excavations at Susa, and we have searched carefully in all our collections for the characteristic wares and forms. It is not easy to recognize these. One problem with recognition is that few Neo-Elamite samples have been excavated and none of these are from Deh Lurān. Another problem is that many types were made from Early Middle Elamite times until the end of the Neo-Elamite phase, and these cannot be used to discriminate Neo-Elamite occupation. Third, many Neo-Elamite (and early Achaemenid) vessels were made from a relatively soft red-brown vegetally tempered ware that does not survive well on the surface of sites. In spite of these difficulties, a number of sites can be dated to this period.

Neo-Elamite occupations are indicated on three sites: two on the Deh Lurān Plain and one to its east (Table 5.5; Fig. 5.5). At 'Ain Khosh (DK-13), just east of the plain, is a complex of two mounds by a spring. On the largest mound (200 m north-south × 150 m, 9.5 m high, and covering about 2.0 ha), Miroschedji (1981b:174, Fig. 62:8–10) found several diagnostic Neo-Elamite sherds. About 17.1 km west-northwest is Patak (DL-35), a former Elamite center abandoned during the later IInd millennium B.C., which produced four diagnostic sherds (Miroschedji 1981b: Fig. 64:1–3; this volume, Fig. 4.24*l*). About 16.7 km farther, the high mound of Gārān (DL-34) produced one diagnostic Neo-Elamite sherd (Fig. 4.24*l*) and several that may be of this period.

We cannot estimate the size for any of these sites in Neo-Elamite times, though none could cover more than a few hectares. The spring of 'Ain Khosh provided water for its settlement. Both Patak and Gārān would have been able to restore their canal (and perhaps *qanāt*) arrangements in place in Earlier Middle Elamite times. Perhaps the most interesting point about the distribution of these three sites is their regular arrangement from east to west, suggesting that a transport route from east to west along the northeastern slopes of the plain had become an important consideration in settlement location and spacing. The lack of sites west of the Mehmeh could be taken to indicate that the threat to Neo-Elamite rule was from the west. However logical this may seem, we must not forget that the western extremity of the plain has not yet been thoroughly surveyed.

The Achaemenid Phase

In 539 B.C., a powerful new military force swept out of the Iranian highlands, absorbed those Elamite polities that had survived the Assyrian and Neo-Babylonian conquest, and overthrew the Dynasty of Babylon. The first Achaemenids were concerned with expanding and consolidating control, but Darius the Ist devoted much of his long reign (522–485 B.C.) to administrative reorganization and royal construction projects.

Later Achaemenid ceramics, well known from excavations at Susa, are found at three sites on the Deh Lurān Plain and one farther east toward Susa. The three relatively large Deh Lurān sites cover a total of about 30 ha (Table 5.6; Fig. 5.6). As few Achaemenid domestic buildings are reported from anywhere in southwestern Iran, we have little basis for estimating the population of these sites.

Of these sites, centrally located Gārān (DL-34) was the largest, the low mounds of Early Dynastic to Middle Elamite date having been incorporated into an extensive lower town on which the gravel footings of large public buildings are visible. The land around this site was watered by a combined *qanāt*-canal system coming from the Dawairij or Āb-Dānān River. We would not be surprised if future study of the many predominantly Sasanian and Islamic sites in this area does not reveal some smaller Achaemenid sites. About 16.8 km to the east-southeast is the small center of Patak (DL-35), watered by the ancient canal from the left bank of the Dawairij. A further 17.1 km east is the site of 'Ain Khosh, described above. About 17.2 km to the northwest of Gārān is the site of Tāfuleh (DL-8), apparently newly founded in this period. While Tāfuleh's fields could have used water from the Mehmeh River, it would have been more reliable to take water from the Ab-i Garm by means of newly proposed Canal DL-331 (see Chapter 6) branching from Canals DL-276A and B, probably unused since the abandonment of Baulah (DL-24) after the Sukkalmaḫḫu phase. Either way, it is an example of canalizing water supplies to sites that are spaced along a major transport route.

In the absence of any excavations whatsoever of Achaemenid remains on the Deh Lurān Plain, there are many questions we cannot answer. The canal systems around Gārān and Patak indicate that crops were grown, but we have no evidence regarding the types of crops or domestic animals. Exploitation of the stone and

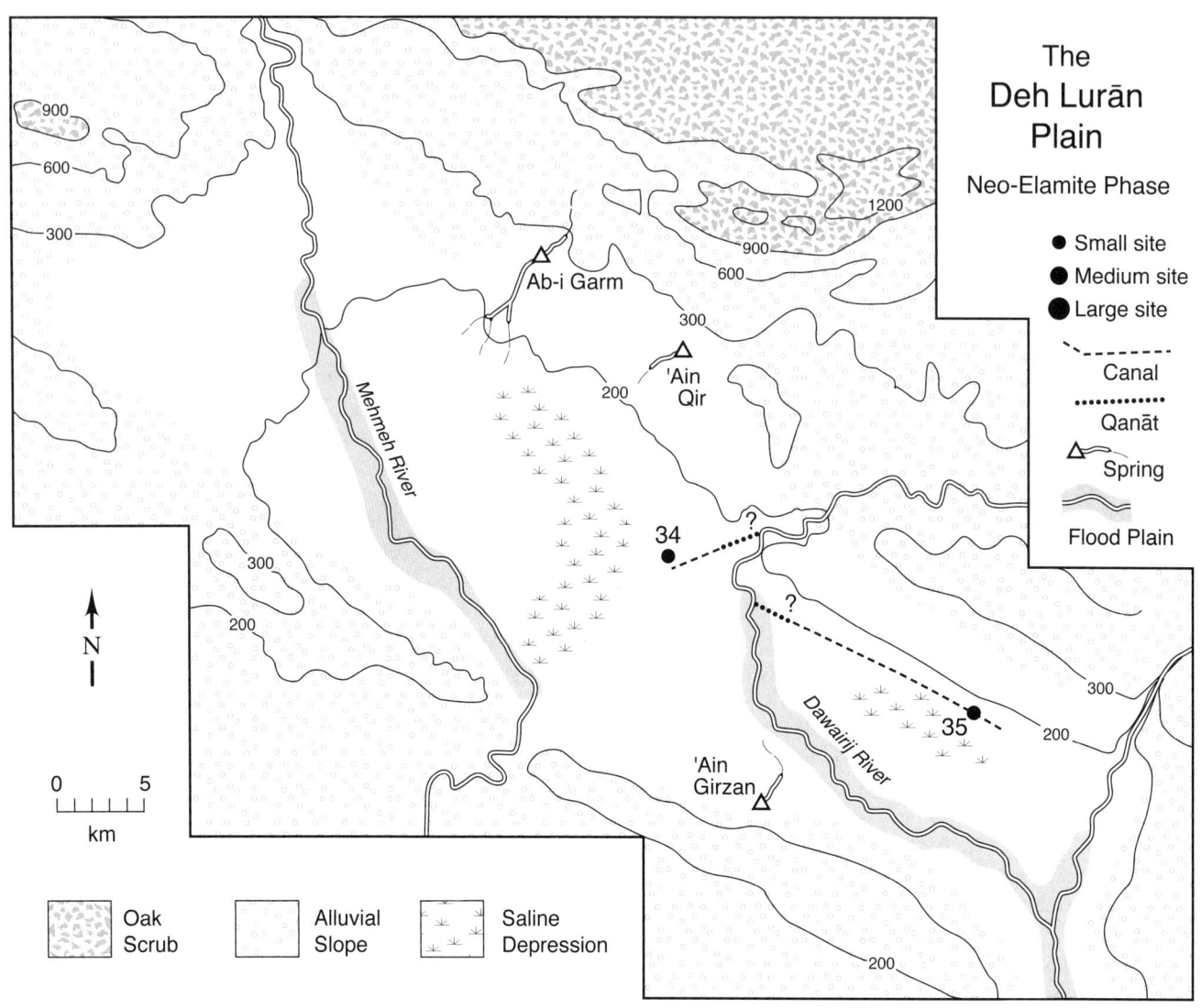

Figure 5.5. Settlements of the Neo-Elamite phase.

Table 5.5. Neo-Elamite settlements.

Site No.	Name	Area (ha)
Centers		
34	Gārān	ca. 12.0
35	Patak	ca. 5.0
–	'Ain Khosh	–

Table 5.6. Achaemenid settlements.

Site No.	Name	Area (ha)
Centers		
8	Tāfuleh	ca. 6.0
34	Gārān	ca. 17.0
35	Patak	ca. 6.5

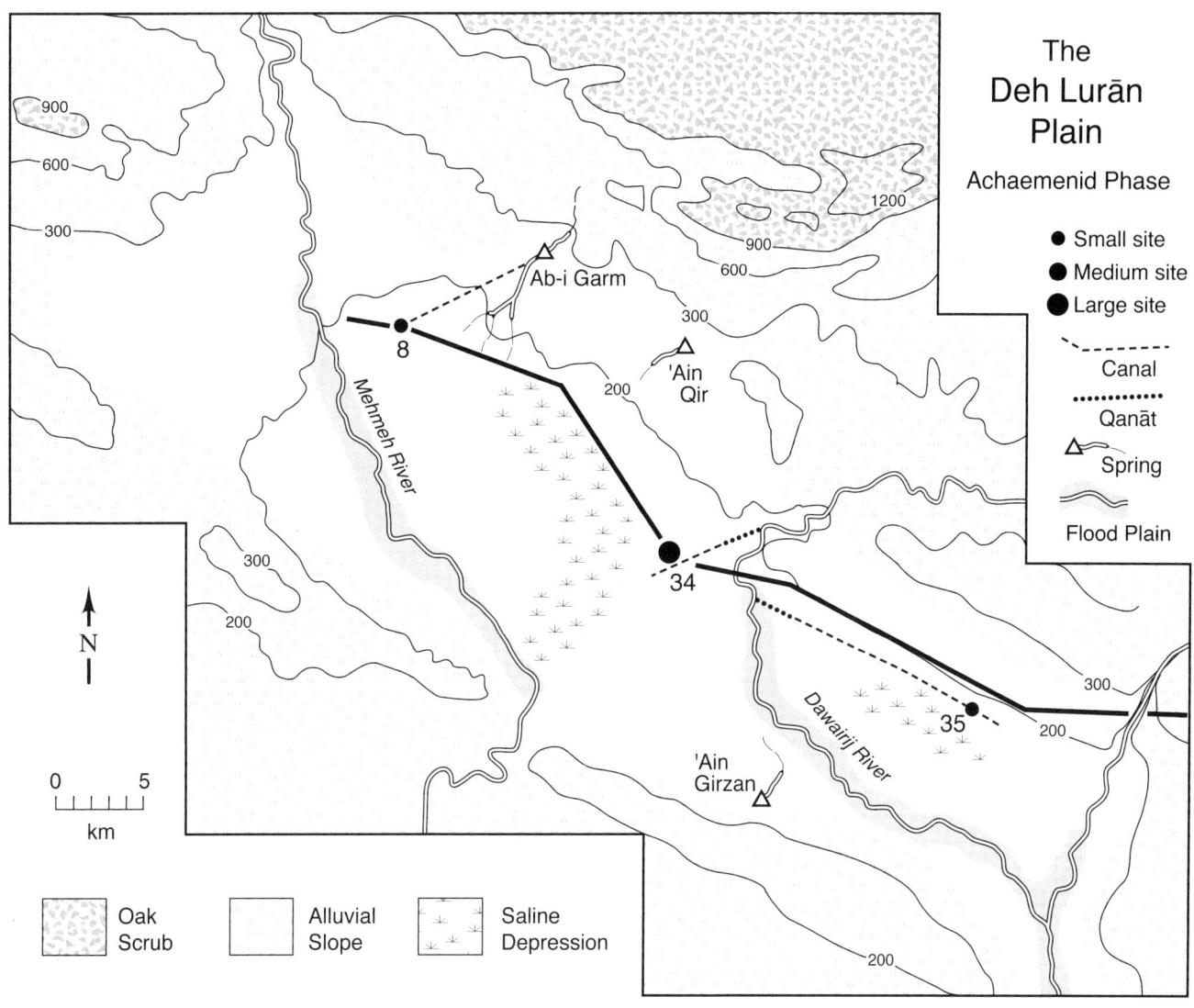

Figure 5.6. Settlements of the Achaemenid phase. (The thick solid line is the proposed Achaemenid Road.)

bitumen resources of the plain must have continued, but we have no evidence. We would like to know whether Elamite domestic patterns continued, or whether there was a change. Obviously, work in layers of both the Achaemenid and pre-Achaemenid phases at Gārān is a high priority when it is finally possible to continue research in the Deh Lurān area.

Chapter 6

Water Management during the Later Early Dynastic, Elamite, and Achaemenid Periods on the Deh Lurān Plain

James A. Neely

Water management and irrigation technology and systems used during the long span of the Later Early Dynastic, Elamite, and Achaemenid periods on the Deh Lurān Plain are not unique. Most have long histories on the Deh Lurān Plain and elsewhere in the Middle East. Those that have been found and recorded in other regions, however, often appear at a much greater scale of size and complexity.

As in our previous and forthcoming studies (Neely 2010b; Neely and Wright 1994; Neely et al., forthcoming), the dating and cultural period assignments of canals and *qanāts* have been only tentatively accomplished based on their association with adjacent sites. It appears that most of the canals we have thus assigned to the Later Early Dynastic, Elamite, and Achaemenid periods were either in continuous use or were refurbishments of canals dating to earlier times. Unfortunately, because our study is based solely on survey, the dimensions (that is, length, width, depth, and slope) of the canals at any specific point in time in the past were impossible to determine. As a result, the dimensions we present here are well-considered estimates that must await excavations for verification. Maintaining an already long-standing practice, the placement of sites was in close proximity to canals, presumably to obtain water for domestic needs. Such placement would also permit water users to take advantage of the flora and fauna that characterized the riparian microenvironment that paralleled the courses of the canals. In association with nine of the ten sites we have assigned to the Late Early Dynastic, Elamite, and Achaemenid phases of occupation, the 1969 survey found each of the three primary canal systems of each of these periods to total over 25 km in length and to have an estimated average width of about 10 m. Each of these primary canals took its water from one of the three major water sources of the Deh Lurān Plain: the Mehmeh River, the Dawairij or Āb-Dānān River, and the Ab-i Garm Springs. In at least two instances, it appears that the Ab-i Garm Springs contributed waters to the canal systems of the two rivers (that is, via Canal DL-331 and Canal DL-330; see below and Appendix B). As noted in Chapter 5, individual springs may also have supplied water to specific sites and their associated nearby fields.

Of the ten sites attributed to these phases of occupation, six sites (DL-20, 34, 35, 41, 54, and 104) were serviced by canals and *qanāts* taking water from the fresh and reliable Dawairij or Āb-Dānān River (with one of these sites, DL-34, perhaps also using Ab-i Garm water); two sites (DL-27 and 32) were supplied with water from the saline and variable Mehmeh River (with one of these sites, DL-32, perhaps also using Ab-i Garm water); while the ninth site (DL-24) evidently made use only of Ab-i Garm water. The tenth site (DL-8) was originally recorded as not being associated with a canal, but we argue below that it did utilize a canal carrying Ab-i Garm water (Table 6.1).

An ex post facto study of site photographs (for example, Plate 7 *upper*) and the settlement pattern suggests that a secondary canal once coursed through the northeast portion of DL-8, but had been obscured by the construction of a narrow-gage railroad. The oil entrepreneur William Knox D'Arcy constructed this railroad from Kut across the western portion of the Deh Lurān Plain to the bitumen spring at 'Ain Qir or Chešme Gir to convey supplies to his oil camp before the First World War. The "bed" of this railroad was not recognized and recorded in 1969 as having paralleled, and in places having been superimposed over, the remnants of what appears to be a much older canal. The railroad bed follows a course that would have been appropriate for a secondary canal branching from Canal DL-276A. The narrow-gage railroad bed and remnants of the proposed canal (DL-331) are visible just to the south of a small structure as it passes through the northern portion of site DL-8 (Plate 7 *upper*).

Considering the settlement pattern, it seems highly probable that DL-8 was supplied water by means of a secondary canal branching from the Ab-i Garm drainage/canal (DL-276A), most likely either about 1.8 km south of where Canals DL-276A and 276B separate from one another (about 800 m north of site DL-7) or just south of site DL-23 (see Appendix B map). The settlement pattern evidence for the existence of this canal is based on the alignment, from northeast to southwest, of sites DL-7, 23, and 11. All three of these sites had an Early Uruk period occupation and very probably could not have existed where they were situated without a domestic water supply provided by a canal. While sites DL-7 and 11 were abandoned after the Early Uruk period, and site DL-23 was abandoned in Late Uruk times (Neely and Wright 1994), we propose that the extant canal (DL-331) that serviced the 2.35 ha site DL-11 (Tepe Garmasi) was refurbished and extended to service the occupants of DL-8 in Achaemenid times. In this same vein, as we will recount in our final volume on the Parthian, Sasanian, and Islamic periods (Neely et al., forthcoming), this canal was subsequently maintained to service the later occupations at sites from DL-7 in the northeast to DL-9, 8, 10, and 220 in the southwest, and may have also supplied water to sites DL-222, 223, and 226 (see Appendix B map) before it emptied into Canal DL-13.

Similarly, an ex post facto consideration of the site of Tenel Ramon (DL-27) strongly suggests that it was serviced by a canal taking water from the Mehmeh River during the Later Early Dynastic and Elamite phases. While no direct evidence was recorded, it seems quite likely that a branch canal from Canal DL-320 led to this site (see Figs. 6.3, 6.4; Appendix B). Furthermore, although we (Neely and Wright 1994:83–84) did not previously recognize the existence of this proposed branch canal, it seems likely that it also serviced at least some of the earlier residents of this site.

Not surprisingly, the waxing and waning of the canal and *qanāt* systems closely parallels that of their associated sites. An overview of these events may be gleaned from Table 6.1 and the settlement pattern synthesis in Chapter 5.

Observations

This study has indicated that there was an apparent shift in the settlement pattern, from a dispersed distribution of settlements with many relatively small sites, to the Elamite/Achaemenid pattern characterized by an apparent population aggregation into a few large sites. Of the ten sites we have defined as having Elamite/Achaemenid occupation, only one (DL-8) was not occupied in earlier periods.

Although all three perennial water sources (that is, the Mehmeh River, the Ab-i Garm Springs, and the Dawairij or Āb-Dānān River) were still in use, there was an apparent shift in emphasis in both settlement location and water use to the eastern portion of the plain. In addition, the eastern sites, on both sides of the Dawairij River or Āb-Dānān River, were by far the largest on the plain during these periods, and were the only occupations to have been serviced by *qanāts*. In general, during the span of time covered by this volume, two different canal regimes existed on the eastern portion of the plain.

The first of these two regimes emerged before Early Dynastic times (Fig. 6.1) and operated until Middle Elamite times (Figs. 6.2–6.4). While there must have been small canals, difficult to detect, sustaining small sites in the west with the saline and variable water of the Mehmeh River, the major sites, such as Musiyān and DL-41, were supplied with the fresher and more predictable waters of the Dawairij or Āb-Dānān River. The newly founded small site of Gārān (DL-34) might have used spring water from the 'Ain Qir or Chešme Gir Spring, but as the site grew it would surely have required water from the Dawairij River. However, the down-cutting of the Deh Lurān rivers (Kirkby 1977:281–83) may have required the development of vented water tunnels (*qanāts*) (Plate 7 *lower*) probably as early as the IInd millennium. As Canal DL-330 may attest, the *qanāts*/canals coming from the Dawairij River were apparently augmented by water from the Ab-i Garm Springs for use at Tepe Gārān (DL-34) and its fields.

The second regime on the eastern portion of the plain perhaps emerged during Neo-Elamite times (Fig. 6.5) and continued through Achaemenid times. Long canals sustained sites carefully spaced along the route from Susa to Der from southeast to northwest across the Deh Lurān Plain (Fig. 6.6). As related above, Tāfuleh (DL-8) was evidently founded along a refurbished earlier canal. Also, the reoccupation of Patak (DL-35) would have required the refurbishing of Canal DL-317C eastward from the already existing but abandoned site DL-41 (Neely and Wright 1994:103–4). In contrast, a *qanāt*-canal system would surely have been necessary to provide water to sustain the increasingly larger settlement of Gārān (DL-34), perhaps during Middle or Neo-Elamite times and certainly by Achaemenid times.

The continuing use or refurbishment and enlargement of older canals as proposed for the Later Early Dynastic, Elamite, and Achaemenid periods should not be considered unusual. As found both prehistorically and historically in several other places throughout the world (for example, the Phoenix [e.g., Masse 1981] and Safford [e.g., Neely 2010c; Neely and Murphy 2008] Basins in the American Southwest; as well as in Peru [e.g., Gelles 1996]; Sri Lanka [e.g., Stanbury 1996]; Sonora, Mexico [e.g., Doolittle 1988]; and Puebla, Mexico [e.g., Neely 2010a; Neely and Rincón Mautner 2004; Woodbury and Neely 1972]), the occupants of the Deh Lurān Plain had refurbished and used canals engineered and excavated by their predecessors.

Save for one instance, we cannot determine what modifications were made to most of the existing earlier canal and *qanāt* systems that supplied water to the Later Early Dynastic/Elamite/Achaemenid phase sites. However, it seems intuitively logical that the growth of these sites during these periods required the enlargement of the canals. They probably also required the excava-

Table 6.1. Proposed formal *qanāts* and canals on the Deh Lurān Plain.*

Canals	Sites	Minimum Length
(M) Later Early Dynastic III Phase (Fig. 6.1)		
Canal DL-121	DL-34 (1.8 ha)	6.5 km
Canals DL-115A, 115B	DL-20 (15.0 ha), DL-104 (2.0 ha) two components, totaling 17.0 ha (mean = 8.50 ha)	13.5 km
Canals DL-317, 317C, 317D	DL-41 (20.0 ha), DL-54 (3.2 ha), DL-35 (1.0 ha) three components, totaling 24.2 ha (mean = 8.1 ha)	16.5 km
(N) Šimaški Elamite Phase (Fig. 6.2)		
Canal DL-121	DL-34 (1.8 ha)	6.5 km
Canals DL-115A, 115B	DL-20 (15.00 ha), DL-104 (0.3 ha) two components, totaling 15.3 ha (mean = 7.6 ha)	13.5 km
Canal DL-317	DL-35	16.5 km
proposed branch from Canal DL-320	DL-27 (1.0 ha)	4.0 km, or 500 m only for branch
(O) Sukkalmaḫḫu Elamite Phase (Fig. 6.3)		
Canal DL-121	DL-34 (1.8 ha)	6.5 km
Canal DL-115A	DL-20 (15.00 ha)	7.7 km
Canals DL-317, 317C	DL-35 (5.0 ha)	16.5 km
proposed branch from Canal DL-320	DL-27 (1.0 ha)	4.0 km, or 500 m only for branch
(P) Earlier Middle Elamite Phase (Fig. 6.4)		
Qanāt and Canal DL-121	DL-34 (12.0 ha)	6.5 km
Qanāt 122 and Canal DL-115B	DL-20 (10.00 ha)	7.7 km
Qanāt and Canals DL-317, 317C	DL-35 (5.0 ha)	16.8 km
proposed branch from Canal DL-320	DL-27 (1.0 ha)	4.0 km, or 500 m only for branch
(Q) Later Middle Elamite Phase		
Qanāt and Canal DL-121	DL-34 (12.0 ha)	6.5 km
(R) Neo-Elamite Phase (Fig. 6.5)		
Qanāt and Canal DL-121	DL-34 (12.0 ha)	6.5 km
Qanāt 117, Canals DL-317C, 317D	DL-35 (5.0 ha)	16.8 km
(S) Achaemenid Phase (Fig. 6.6)		
Qanāt and Canal DL-121	DL-34 (17.0 ha)	6.5 km
Qanāt 117, Canals DL-317C, 317D	DL-35 (6.5 ha)	16.8 km
proposed Canal DL-331	DL-8 (ca. 6.0 ha)	6.0 km

*This table presents a listing of the distribution, by phase, of the identified formal (i.e., fully human engineered and excavated) canals and *qanāts*, as well as the dated components of the eight sites found in association with well-defined formal water management systems. These data are accompanied by estimates of the area of occupation of each site for each of the phases, as well as the total and mean occupation area for each phase based on the estimates presented in Chapters 4 and 5. We have also included the minimum length of each of these formal canals and *qanāts* for each of its phases of use, as indicated by their traces found on aerial photos and site locations. The letter designations given to each of the cultural phases (i.e., *M–S*) are continuations of those applied to the cultural phases considered in Neely and Wright (1994:196–98).

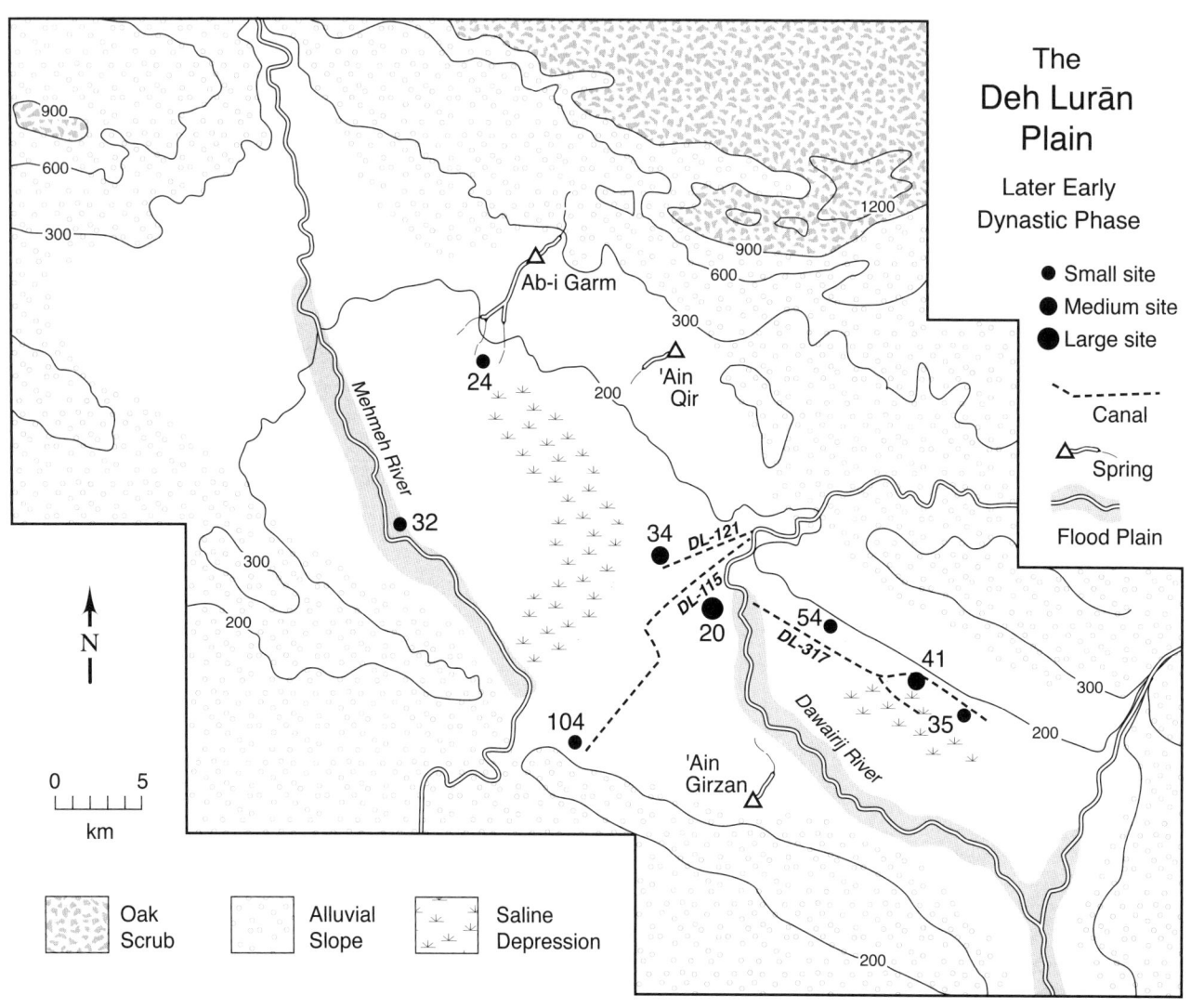

Figure 6.1. Later Early Dynastic phase canals.

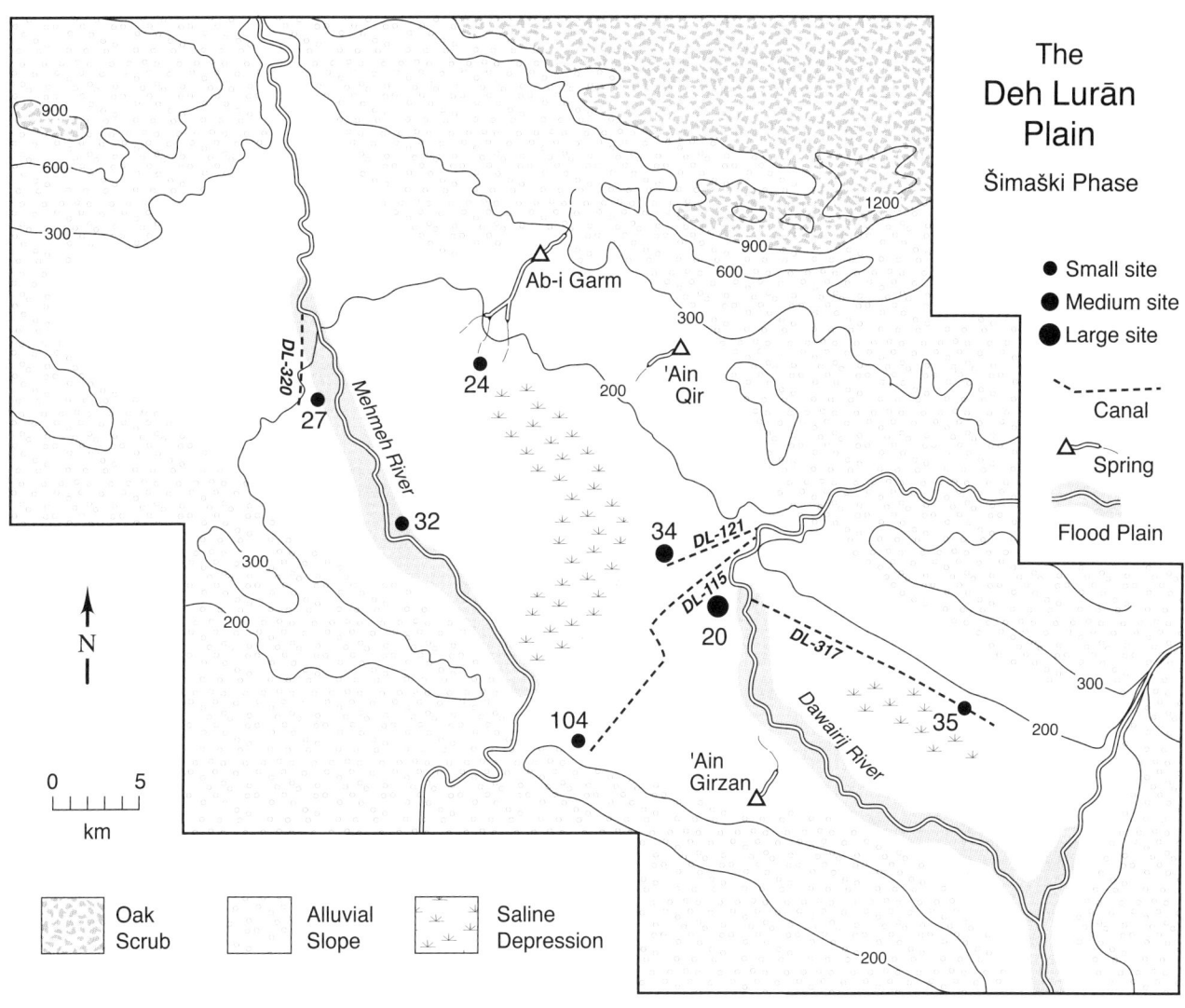

Figure 6.2. Šimaški phase canals.

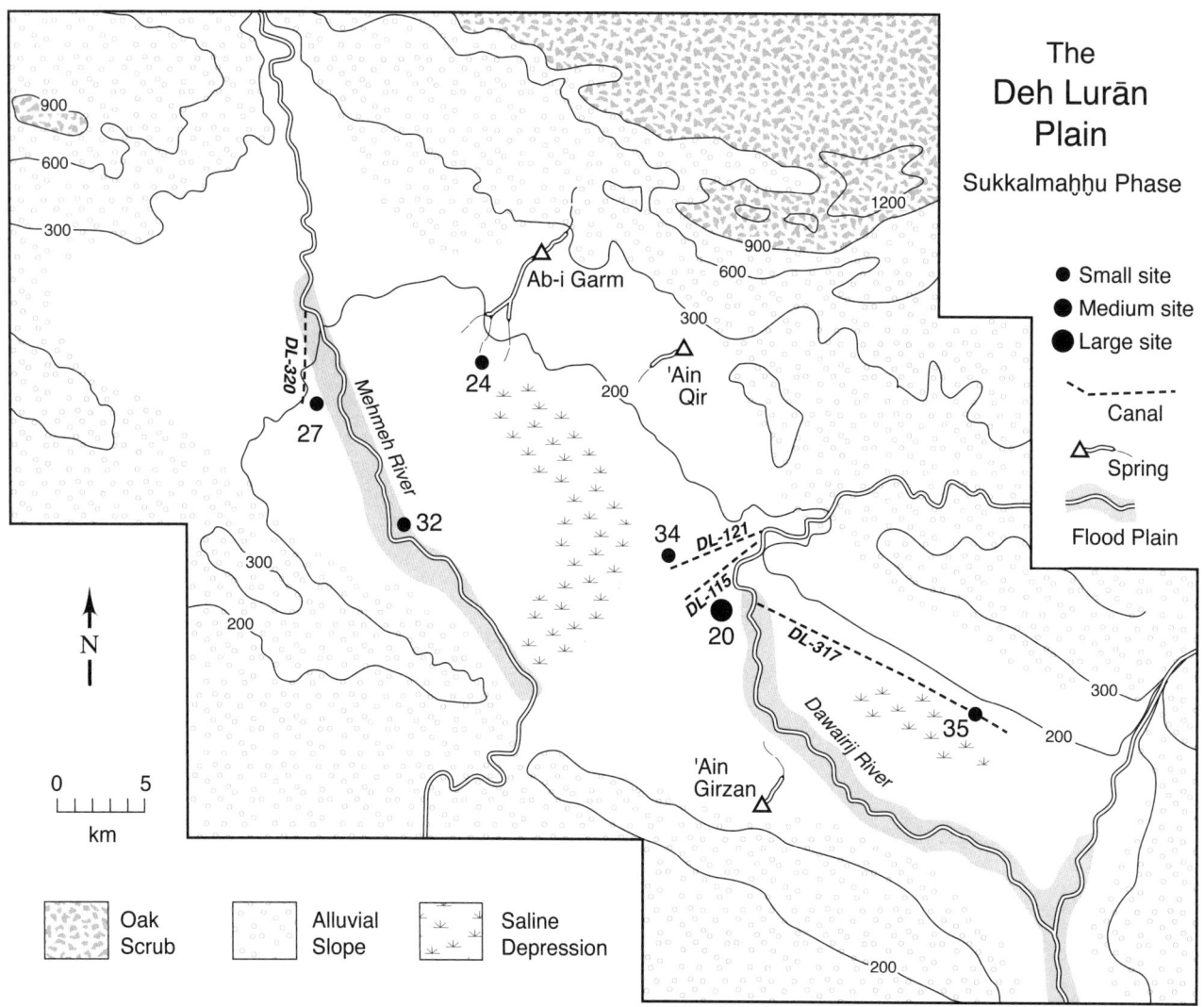

Figure 6.3. Sukkalmaḫḫu phase canals.

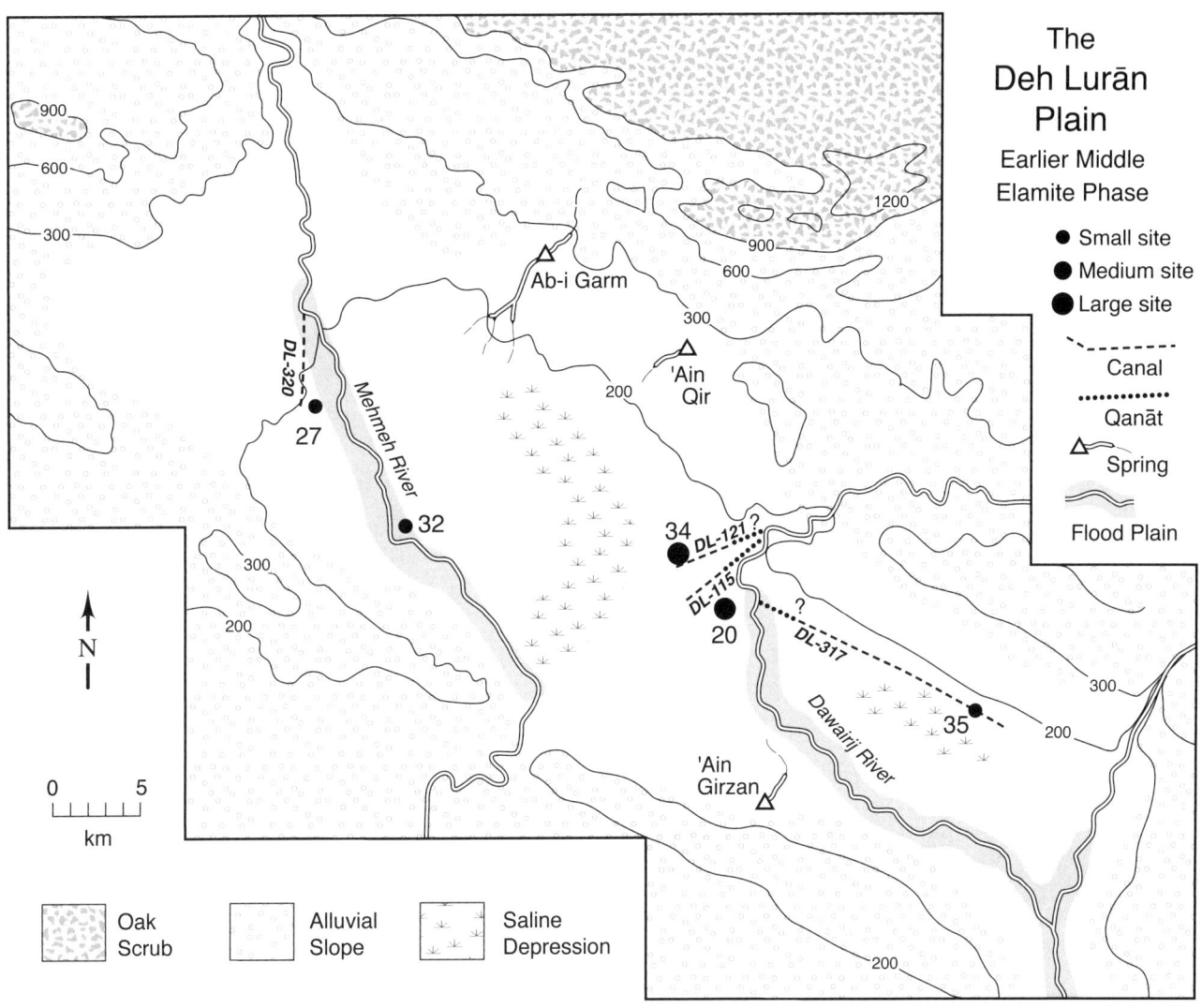

Figure 6.4. Earlier Middle Elamite phase canals and possible *qanāts*.

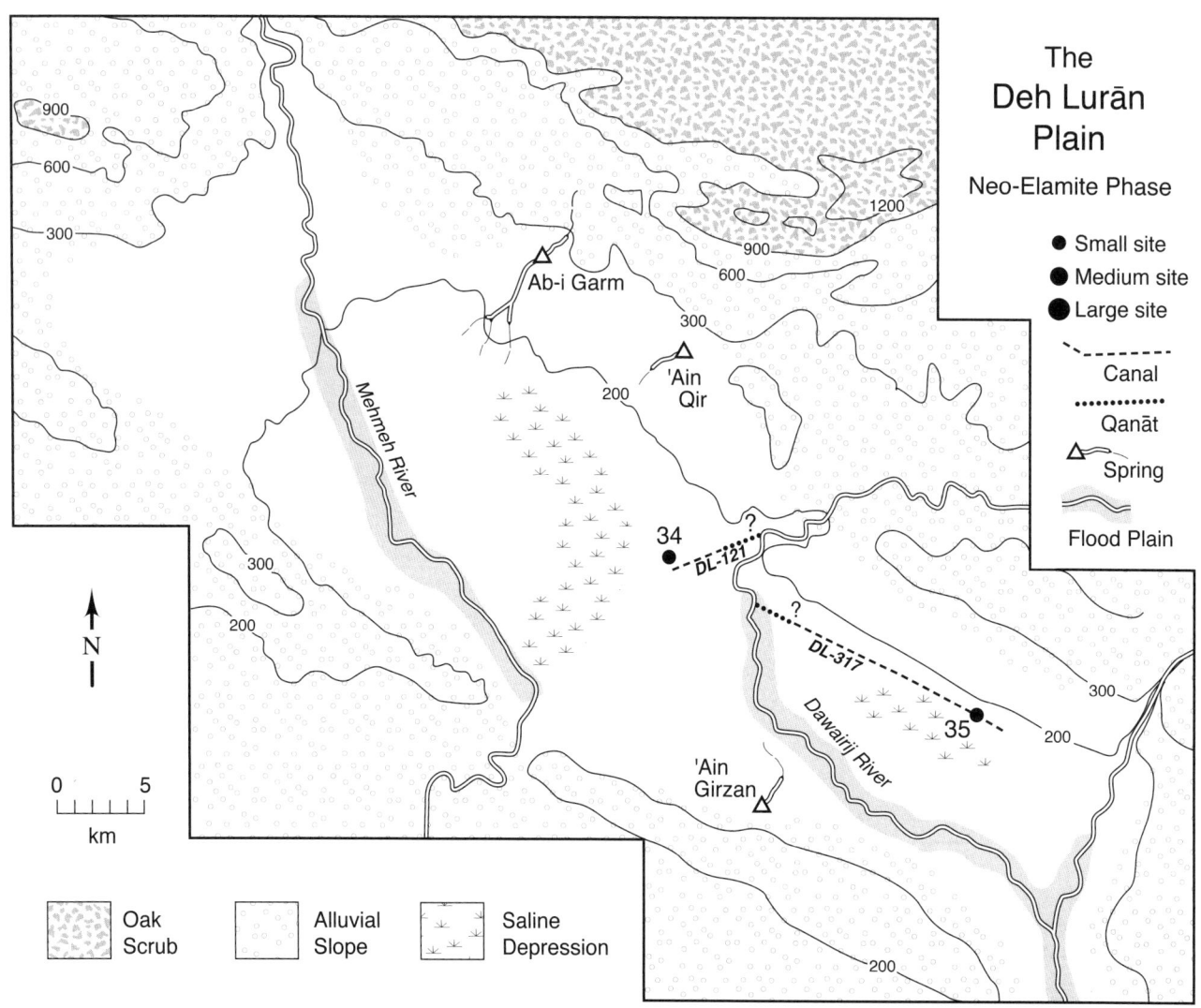

Figure 6.5. Neo-Elamite phase canals and possible *qanāts*.

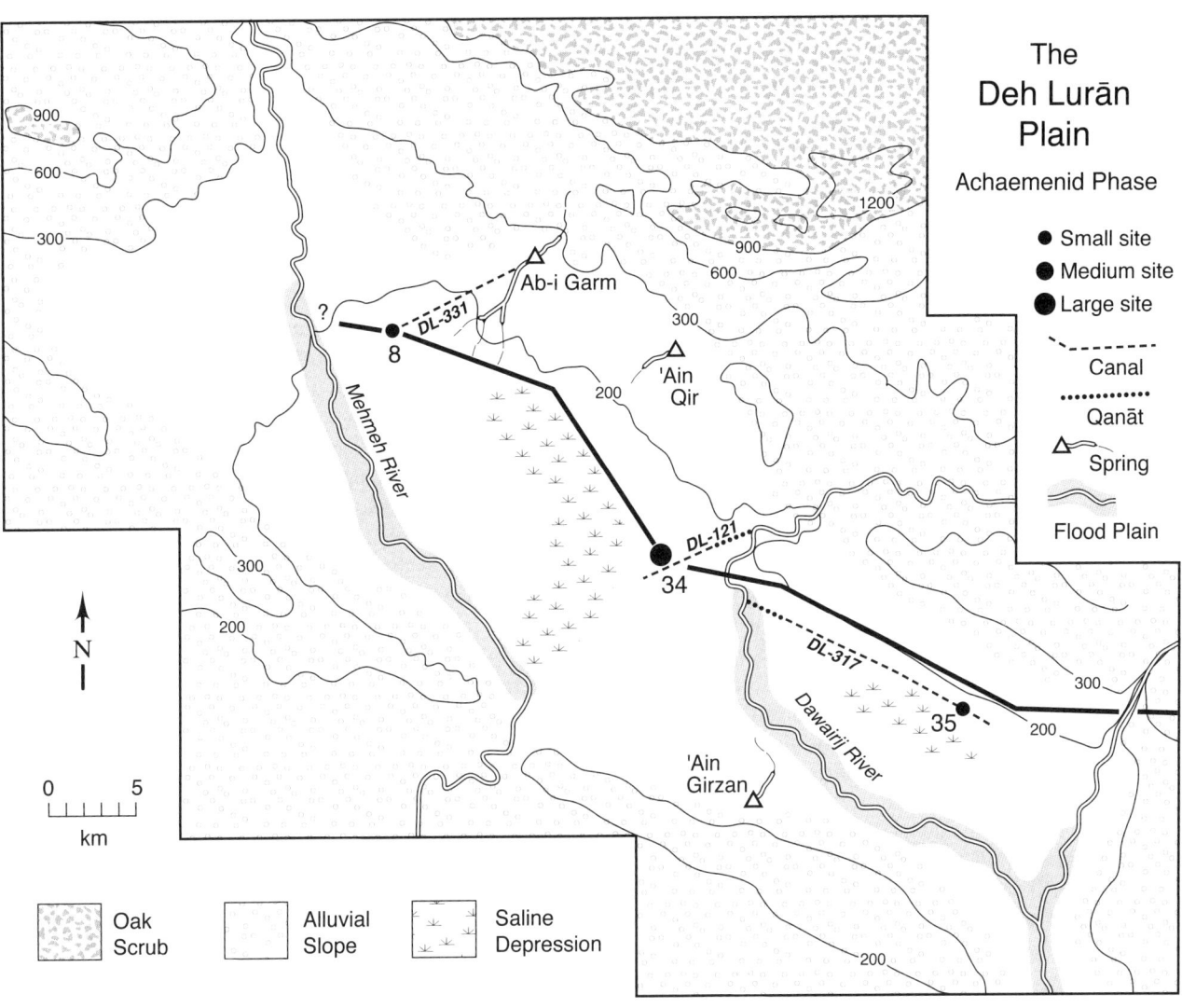

Figure 6.6. Achaemenid phase canals, *qanāts*, and roads. (The thick solid line is the proposed Achaemenid Road.)

tion or augmentation of some of the *qanāt* segments of the canal systems (that is, *qanāt* DL-121 for site DL-34; *qanāt* DL-122 connected to Canal 115 for sites DL-20 and DL-104; *qanāt* 117 for site DL-54; and *qanāt* DL-317 for sites DL-41 and DL-35). The one instance where a probable modification was observed is the apparent double offtake affecting the DL-317 canal. In this case, we documented both a canal offtake and a *qanāt* supplying water to the canal (see Appendix B). While we cannot verify the sequence of construction without further fieldwork, it seems highly probable that the *qanāt*, located about 100 m south of the canal offtake, was constructed at a later time when the downcutting of the river channel rendered the original offtake useless.

Reevaluating our previous statement (Neely and Wright 1994:199–200) regarding their earlier origins, it seems more likely that *qanāts* may have made their appearance for the first time during the IInd millenium rather than the IIIrd millenium. It is probable that *qanāts* replaced the upstream portions of some canals—that *qanāts* were excavated adjacent to older canals when the waters of the Dawairij or Āb-Dānān and Mehmeh Rivers cut so deeply into their channels that it was either no longer feasible or efficient to maintain the upstream portions of the canal system. This probability is reinforced by the observations of Kirkby (1977; Kirkby and Kirkby 1969) that the down-cutting process of the Deh Lurān riverbeds began by the IInd millennium B.C., and that this process undoubtedly affected the locations of canal offtakes. Our observations also indicate that some *qanāts* were engineered and excavated to intersect functionally with an existing earlier canal channel at the point where the *qanāt*'s nearly horizontal tunnel broke to the surface of the sloping landscape. New methods, such as the dating of soil sampled from the sediment around the *qanāt* vents with Optically Stimulated Luminescence (OSL), could resolve the issue of the first use of *qanāts*.

In addition, if not initiated earlier (Neely and Wright 1994:200), the above-mentioned down-cutting process may have precipitated an ingenious use of *qanāt* technology during these periods. This modification was to obtain seepage waters from the Mehmeh and Dawairij or Āb-Dānān Rivers rather than to tap underground aquifers as traditional *qanāt* technology does. The vented tunnels did not take water directly from the rivers, but ran parallel to them at distances from ten to fifty meters for stretches of two hundred meters to nearly two kilometers before turning away from the riverbanks and continuing as canals. Water was obtained from the rivers as it percolated through the soil and rock of the channel banks into the *qanāt*. This variant technology may well have been contrived to serve four purposes. First, it greatly reduced the amount of silt carried into the systems—a problem that must have been enormous in canal systems that took water directly from the Mehmeh and Dawairij Rivers by means of diversion dams or weirs. Second, the percolation of the water through the riverbanks filtered out vegetal matter and minerals in suspension. Third, it is conceivable that this filtering process resulted in the deposition of dissolved minerals in solution through evaporation as hardened calcium carbonate (*caliche*) layers at the top of the aqueous zone. Fourth, although the inhabitants of Deh Luran were undoubtedly unaware of it, the filtration of water through riverbanks has been documented as serving to remove microbes and other pollutants (e.g., heavy metals) from water supplies (Hubbs 2006; Ray et al. 2002). Three of these processes reduced the frequency of the need to clean the systems, and slowed the soil salinization that probably was gradually affecting crop production. The fourth benefited the health of the inhabitants of the now large and densely populated communities.

Our survey did not discover specific evidence of dry-farming or irrigated fields dating to the Later Early Dynastic, Elamite, or Achaemenid periods. This is not surprising considering the ongoing alluviation of the plain (Kirkby 1977; Kirkby and Kirkby 1969), as well as the preservation of ancient fields on the Deh Lurān Plain dating to later periods (Neely 2010b; Neely et al., forthcoming). However, the use of appropriate new and/ or specialized survey and excavation techniques have revealed early fields in other parts of the Middle East (Wilkinson 2003), Mexico (e.g., Neely and Rincón Mautner 2004; Neely et al. 1990; O'Brien et al. 1982), and the American Southwest (e.g., Diehl 2009; Herr 2009a, 2009b; Mabry 2008; Neely 1995, 2005, 2010c). Thus, the success of concentrated future efforts in the search for Elamite/Achaemenid fields bodes well.

In the first Deh Lurān survey volume (Neely and Wright 1994), it was possible to make relatively precise estimates of prehistoric village populations and to determine that there was a close correlation between canal length and the number of people living in villages along the canal, with a mean of 17 people for each kilometer of canal. However, with the growth of the town at Musiyān this relationship broke down, and during Early Dynastic I-II times the estimated number of people per kilometer of canal that provided water to the Musiyān area rose to 300 (Neely and Wright 1994:212–14). During the Later Early Dynastic, Elamite, and Achaemenid phases covered in the present volume there are few villages, and the estimated number of people per kilometer of canal feeding a town varied from 230 to 550. This is certainly enough labor to maintain these canals, but were there fields sufficient to feed the people?

Although it was not possible for us to determine the extent and locations of the cultivated fields during the Later Early Dynastic, Elamite, and Achaemenid phases, one would suspect that they did not contract from the previous period. With the lesser number of larger sites of these phases on the landscape, one must suspect that the Deh Lurān farmers had to walk a further distance daily to their fields, or utilize small field camps with perishable structures, to permit in-field residence for several days without the need of commuting.

As briefly mentioned above, the many canals of the Deh Lurān Plain very likely generated riparian microenvironments along their courses. Such riparian zones would have provided a ready supply of edible and useful plants, and would have been a haven for animals that would have found the zone a source of both water and food prior to their kill or capture for human use. This riparian microenvironment phenomenon along canals was observed on the Deh Lurān Plain (Plate 6 *lower*) and in other locales in the Middle East in 1969, as well as more recently in the American Southwest (Neely 2005, 2010c; Neely and Murphy 2008), along the courses of the canals in the Tehuacán Valley of southern Mexico (Neely 2010a; Neely and Rincón Mautner 2004), and elsewhere.

Chapter 7

Textual Documentation of the Deh Lurān Plain: 2550–325 B.C.

Piotr Michalowski, Pierre de Miroschedji, and Henry T. Wright

During the span of time covered by this study, archaeology is only one of several sources of information about the Deh Lurān Plain. We have also an episodic, fragmentary and often delphic textual record. Unfortunately, many texts remain unpublished. Fortunately, the published texts have been the object of generations of thoughtful research and text criticism by scholars with increasingly skilled understandings of the long-dead languages and of the nature of textual records in general. This chapter is merely an introduction to these sources for the benefit of archaeologists who may not fully realize their great potential value.

Nonetheless, there are many possible pitfalls. The text sample from each historical period is biased in different ways, because of both scribal practice and the vagaries of preservation and discovery. These issues will be discussed separately for each period. A general problem is the representation and interpretation of place names. One cannot rely on the transliterations and interpretations provided by various scholars, but must look at copies or images of the text themselves. One example will suffice. In the following section, a place name transcribed as URU×A^{ki} is discussed. In texts of the mid-third millennium B.C. Early Dynastic period, this is written as the sign for "A" inside the sign for "URU," the latter having the general meaning of "town" or "city," followed by the determinative for a place "k i." If the determinative "k i" is not present, the composite sign URU×A is probably not a place name, and indeed in several Sargonid texts, it appears to be part of a personal name (e.g., Thureau-Dangin 1910: Texts 1075, 1174). In later mentions during the IIIrd Dynasty of Ur, many interesting texts mention a place the name of which is transliterated as ÚRU×Ú^{ki} that has been equated to URU×A^{ki} and read "U r u a." However, it is now clear that this is to be read "Ú r u m^{ki}," a town near K i š in the northwestern alluvium of Mesopotamia unrelated to URU×A^{ki} (Michalowski, pers. comm.; Sigrist 1988: Text 927; Steinkeller 1980). Similarly, a place the name of which is written URU×GU (Dahl 2007:18), where there was an agricultural estate of one of the queens of the ruler Šulgi, has been read "U r u a," but the infixed sign "GU" is very different from "A" and though we do not know where it was, it is probably not the same as URU×A^{ki}. On the other hand, a common Ur III place name transliterated as U.URU×A.A^{ki}, though only in messenger texts from G i r s u, is probably the same as URU×A^{ki} because both forms are often mentioned in conjunction with Susa and other towns in what is today southwestern Iran (Steinkeller 1982:244–45, n. 26). The initial "U" and the terminal "A" probably served to tell the reader something about the pronunciation of the name. A Sumerian literary composition of the early IInd millennium seems to show that URU×A^{ki} was phonetically rendered as "*Arawa*" (Van Dijk 1978; Frayne 1992), but this is a pronunciation as heard by speakers of a Semitic language. We do not know the language of URU×A^{ki}—it need not have been Akkadian or Sumerian or Elamite—and so we will continue to use the graphemic representation until we have a contemporary and local documentation. In sum, it is important to confirm the actual signs used in cuneiform texts, and because we often have only a transliteration not collated with the original text, this chapter cannot be a complete assessment of the published textual evidence.

The Mid–Late Third Millennium B.C. on the Deh Lurān Plain
(by Piotr Michalowski and Henry T. Wright)

Land and People in Early Elam

Through more than five millennia of record keeping in the Near East, Deh Lurān's geography and history are never well or richly recorded. Even in the nineteenth and twentieth centuries the documentation available to scholars is poor. However, during a few spans of time, often with the help of a few conjectures that are not yet proven, we can evaluate and expand our archaeological knowledge with insights from documents. One such period is that from Later Early Dynastic times to the end of the IIIrd Dynasty of Ur, approximately 2550 to 2000 B.C. Our knowledge of the period comes largely from many accounts dealing with labor and goods, some texts conveying royal propaganda, rare literary works, a few letters and legal decisions, and a few word lists. All these were kept by scribes trained in the Sumerian and Akkadian languages and thus in the perspective of Lower Mesopotamia. We are fortunate that in recent decades a diversity of philologists have put much effort into understanding the geographical and ethnic terminologies of these texts (Vallat 1980, 1993; Stolper 1984; Selz 1991; Frayne 1992; Steinkeller 2007).

The logographic sign that represents Elam /elam (NIM)/ occurs in word lists and as an element in lists of personal and divine names as early as 3100 B.C. (Potts 1999:87). Lists aside, the earliest mention of the region today known as southwestern Iran is a supposed attack on Elam by Enmebaragesi (Enišibbaragesi), the lugal or ruler of Kiš on the northern alluvium of Mesopotamia ca. 2650 B.C. Enmebaragesi may be attested by contemporary documents (Postgate 1992:28–30) although his historicity has been questioned of late (Michalowski 2003). However, the reference to the attack on Elam is in the Sumerian King List (Jacobsen 1939) composed no later than 2100 B.C. (Rowton 1960; Steinkeller 2003) and is probably no more than an historical anecdote. There is, however, a IInd millennium copy of a text by an otherwise unknown IIIrd millennium king of Kiš by the name of Enna-il, who claimed to have defeated Elam (Goetze 1961:107–9). Later, Eannatum, the ruler of the regional state of Lagaš about 2460 B.C. claims in his own texts to have fought with Elamite armies (Selz 1991). He lists a number of towns in Elam as his opponents, among them URU×A^{ki}, discussed below. In addition to such political claims, people appear in Lagaš texts with the designation of Elamites /elam/, some of them having names linguistically recognizable as Elamite (Zadok 1994). However, Michalowski (2008:110–13) has argued that the term, in Ur III times at least, was not an ethnic term, but was used for a kind of military profession, more precisely for bodyguards of messengers and diplomats coming from the east. Mesopotamians at this time, however, seem to view the highland areas to the northeast of lower Mesopotamia as the region of Elam, but used different terms for peoples of the foothill valleys (Vallat 1980; Michalowski 2008:112).

Two places mentioned as mid-third millennium localities in later texts—thought to be copies of mid-third millennium compositions (though rarely in contemporary mid-third millennium texts)—are Susa and Awan. Both appear as political centers in later texts. Susa, the site of almost a century of research by French archaeologists, has a rich archaeological and textual record and its location and character are well known. During the later IIIrd millennium, Susa was an urban center covering at least 40 ha dominating many smaller settlements on the Susiana Plain. *Maništusu* and *Naram-Sin*, the third (or possibly second) and fourth Sargonid rulers, had governors at Susa and the latter dedicated a temple there (Stolper 1984; Potts 1999:106–8). Awan, however, is not yet located. The Sumerian King List (see above) states that Awan attacked Ur in southern Sumer about 2450 B.C. *Rimuš*, the second (or third) Sargonid ruler, in a late copy of one of his texts, claimed to have won a battle by the *Qablitum* River, an otherwise unknown watercourse between Susa and Awan, at about 2300 B.C. (Gelb and Kienast 1990:207ff). The names of Susa, Awan, and several other places were, however, used as elements in the titles of offices from Sargonid times onward, and these are taken to be the designations of polities or districts of Elam. Vallat (1993) suggests that Awan covered much of the central Zagros. Potts (1999:97–98) argues for a more limited area in Luristan, including both the Pusht-i Kuh (meaning "behind the mountain," the mountain being the Kabir Kuh, from the perspective of the Iranian plateau, and including Deh Lurān) and the Pish-i Kuh (meaning "in front of the mountain"). Michalowski (2008), in light of newly available texts, agues that Awan was a political unit located northeast of Susa, perhaps controlling parts of the northeastern Susiana and the central Zagros. None of the arguments are certain. Thus, while the location of the town of Susa and the surrounding lowland district of Susiana are firmly identified, the location of Awan—whether it was town, a district, or a more ephemeral political coalition—is not.

Late in the IIIrd millennium, political conditions changed and geographical knowledge expanded. Two new geographical names, first "Anšan" and then "Šimaški," became important. Thanks to clues in Achaemenid texts from Persepolis and to the excavations of the University of Pennsylvania team during the 1970s, the town of Anšan can confidently be located at Tal-i Malyan, the Kur River Basin in central Fars on the plateau of southern Iran (Hansman 1972; Reiner 1973; Sumner 1974). During the IIIrd millennium it was a large urban center with a wall enclosing about 150 ha. Unfortunately, layers of the Late IIIrd millennium Early Kaftari phase have not been extensively excavated, and the few documents found have not yet been critically studied. Texts from Mesopotamia, however, attest to frequent interaction with Anšan. Šimaški is thought to have been located in the central Zagros northwest of Anšan (Stolper 1982; Steinkeller 2007; Michalowski, in press), but neither Šimaški nor any of the other major towns in the district of Šimaški mentioned in Mesopotamian texts have been precisely located.

*Deh Lurān in the Time of the Regional States
and the First Empires*

We are indebted to Piotr Steinkeller (1982) for framing the argument that the town with its name written URU×A^ki was to the northwest of Susa. Steinkeller reasons that scribes writing in lower Mesopotamia refer to URU×A^ki as /sag-kul.elam^ki [ma]/ or "the bolt [of] Elam," just as they referred to the town of Ḫuḫnuri, located somewhere in the foothills of the southern Zagros in eastern Khuzestan, as "the bolt of Anšan," a region still farther east in southern Iran. Since the traditional land route from lower Mesopotamia to Susa goes via *Der* southeastward along the foot of the mountains into Susiana, approximately modern Khuzestan, URU×A^ki should be somewhere between *Der*, near modern Badra in Iraq, and Susa, in central Khuzestan in Iran. The Deh Lurān Plain is the only large tract of agricultural land on this route. Therefore, URU×A^ki should have been on the Deh Lurān Plain. Furthermore, Steinkeller (1982) noted that URU×A^ki is associated with bitumen. Indeed, there is a large bitumen spring, 'Ain Qir or Chešme Gir, on the north fringe of the Deh Lurān Plain close to Musiyān (Marschner et al. 1978; Wright ed. 1981), used from earliest Neolithic times (Gregg et al. 2007) until the early twentieth century. Realizing that Tepe Musiyān is the only large site occupied during the time of the Early Dynastic period, the Sargonid Dynasty, and the Third Dynasty of Ur when URU×A^ki is mentioned in a diversity of cuneiform texts, Elizabeth Carter (in Carter and Stolper 1984:212, n. 275) explicitly identified Musiyān with URU×A^ki. This same evidence led Douglas Frayne (1992:71) to agree that the toponym in question was located in the Deh Lurān Plain. A few years later he was more exact, stating that "the research of this author would strongly suggest that the city is to be placed at the archaeological site of Tell Farukhābād" (Frayne 1997:233). This specific ascription is, however, unlikely, since Farukhābād was quite small in this period (Wright ed. 1981:97–99, and this volume).

The only other geographically possible route from *Der* to Susa is one going southeastward along the south side of the Jebel Hamrin just inside modern Iraq. Because of political circumstances, this terrain is inaccessible to archaeological surveyors. However, satellite images show no large mounds on this topographically possible route, and there is no later tradition of its use.

During most of the Early Dynastic period, Musiyān (DL-20) was the predominant central town of the Deh Lurān Plain. Settlement is focused on the west end of the plain. The archaeological evidence summarized in Chapter 5 indicates that during the earlier Early Dynastic period, Deh Lurān was the location of a small and compact hierarchically organized polity centered on Tepe Musiyān, with several subsidiary towns (e.g., Tepe Baulah [DL-24] and Tepe Farukhābād [DL-32]) and a number of dependent villages. The Deh Lurān area exported locally extracted bitumen and chert and may have exported other stones, wood, and animal products. It had a settled population of about 5000 people, about 3000 of whom lived at Musiyān. We have interpreted this as a small state prospering as a result of its position on an important trade route. At the end of the period, the western plain was largely abandoned and a village on the east end of the plain (DL-41) was expanded and fortified.

If Musiyān was URU×A^ki, these archaeological inferences can be confirmed and expanded. Early during this time URU×A^ki appears as the capital of an independent polity. About 2460 B.C., Eannatum, the énsi or ruler of Lagaš, a large regional polity in the far south of Lower Mesopotamia centered about 160 km southwest of Deh Lurān, campaigned against coalitions involving URU×A^ki. The Eannatum inscriptions document claims of victories against two large foreign coalitions that threatened his borders, one northern that included the northern Babylonian polities of Kiš and Akšak, as well as Mari, situated further north on the Euphrates. The other one, defeated at a Lagašite canal, came from the east and consisted of Elam, Subir, and Urua (URU×A^ki; Frayne 1991:148 vi 17–20). Elam and Subir are broad geographical terms that referred to the southwestern and northeastern highland areas on the borders of Mesopotamia, suggesting that in the conceptions of Eannatum's scribes, Urua was more than just a city, and at the same time testifies to its importance at the time. Particularly informative is a statement that "the énsi of URU×A^ki put the šu-nir [the standard] beside the vanguard of his city, waiting for the troops of Eannatum" (Sollberger 1956: Eannatum Brick A 1 III 10 to 1 IV 2). Szarzyńska (1996) has shown that standards placed or carried on a pole have represented institutions and polities from at least 3200 B.C. The position of the ruler of URU×A^ki in this battle, wherever it took place, indicates the importance of the polity in regional coalitions. Relations, however, were sometimes peaceful. For example, during the reign of Entemena of Lagaš, about 2450 B.C., Dudu, a sanga (a head temple administrator) of the god Ningirsu, imported "bituminous limestone" from URU×A.A^ki for a door peg (Sollberger 1956: Entemena Text 76; Frayne 1991:232–33). A text concerned with trade with Elam found at Girsu, a major town in the Lagaš polity, records delivery of 120 bushels of barley and one mana of tin to the énsi of URU×A^ki (Selz 1989:539: Nikolski Text 310 viii 3–9). The text is not dated, but the script indicates it was written very late in the Early Dynastic period, probably during the reign of Lugalanda or UruKAgina about 2400 B.C., long after the time of Eannatum. In an undated Early Dynastic text, five mana of silver are brought from URU×A^ki (Foxvog 1980:68, 71: Text UCLM 9-1798: v 66–vi 73). This would involve an overland journey of at least 340 km from URU×A^ki to a point on the Tigris south of *Der*, perhaps two weeks travel, followed by a trip by canal barge to Adab. Why so much silver would be taken from a small town in the foothills to a large city on the alluvium is unclear, and Foxvog (1980:74–75) suggests that this was in fact not the place named URU×A^ki in Elam, but another. While the destruction wrought by Eannatum (and perhaps the campaigns of other rulers not so well documented) may have caused the movement of population eastwards on the Deh Lurān

Plain, there is no indication in the texts of the establishment of the major new town marked today by DL-41. This simply emphasizes the sad fact that the published later Early Dynastic texts are relatively few, and primarily from one city: Girsu, the older center in the Lagaš polity. While we know much about Girsu and Lagaš, we do not have enough texts to characterize other polities, including URU×A^{ki}.

During the later IIIrd millennium, the archaeological period termed "Šimaški" in preceding chapters probably covers the time from the middle of the Sargonid Dynasty until after the fall of the Third Dynasty of Ur (Ur III). Musiyān was again the sole large center on the plain, DL-41 having been abandoned. On the plain to the west was the small center of Baulah (DL-24), Farukhābād (DL-34) having diminished to the size of a hamlet. Both Musiyān and Baulah had a few subsidiary smaller villages and hamlets. Unfortunately, the archaeological evidence indicates stylistic relations with both Lower Mesopotamia proper and Susiana and there is little other evidence of exchange of subsistence or craft goods in our samples from the hamlet of Farukhābād. The aggregate population is estimated at 4600 people. Fortunately the available documents complement and greatly expand the archaeological evidence. The documents indicate that under the Sargonid Dynasty, URU×A^{ki} was conquered and apparently incorporated into the imperial polity. Sargon, the founder of the dynasty about 2350 B.C., named one of his early years in office after his destruction of URU×A^{ki} (Gelb and Kienast 1990:49–50). Subsequently Elam was brought under enduring control. A Sargonid énsi, which at this time means a governor rather than an independent ruler, was in place at Susa during the reigns of Sargon's son *Maništusu* around 2300 B.C. and, following the suppression of rebellion in "all of Elam," his grandson *Naram-Sin* (Stolper 1984:13–14; Potts 1999:107–8) also appointed governors. *Naram-Sin*'s successor, *Šarkališarri*, was fighting Elamites in Mesopotamia proper around 2250 B.C. and it seems likely that Sargonid control of Susa was lost during his reign. URU×A^{ki} appears rarely in these political texts, and may have been ruled as an adjunct of Susa. One indication of the close relation between URU×A^{ki} and Susa is an undated Sargonid economic text from Susa that records the delivery of 195 bushels of barley from URU×A^{ki} to Susa (Scheil 1913: Text 21). If URU×A^{ki} was in the Deh Lurān area, this is an overland trip of four or five days from Susa involving about 100 pack animals. During the interregnum after the end of Sargonid control, when Gutian people from the Zagros occupied parts of Lower Mesopotamia, there is no reference to URU×A^{ki}.

Under the Third Dynasty of Ur, URU×A^{ki} was closely integrated into the governmental and economic structure of the empire. The founder of the dynasty, Ur-Namma, apparently defeated the ruler of Susa, Puzur-Inšušinak (Potts 1999:122–25; Wilke 1987:108), and drove the Elamite troops from towns in the northern alluvium, but there are no explicit mentions of the conquest of URU×A^{ki}, Susa, Awan, or any place in Elam in Ur-Namma's known inscriptions. Puzur-Inšušinak is known from a number of dedicatory inscriptions at Susa and is listed as the last king of the Awan Dynasty in a text of ca. 1800 B.C. from Susa (Stolper 1984:12, 15; Potts 1999:122). However, Ur III governors holding the title "énsi of Susa" were appointed from the middle years of Ur-Namma's son and successor Šulgi about 2080 B.C. until the later years of Šu-Sin about 2030 B.C. Battles were fought and diplomatic marriages were made with rulers farther north and east in the Zagros (Stolper 1984:16–19). Various government employees received rations for going to and from URU×A^{ki}. To select only a few examples, among these are people termed "sukkal," probably "emissaries" in such cases (Jones and Snyder 1961:124, Text 212; Sigrist 1990: Texts 74:8, 165:3); people termed "lú-kaš$_4$," probably "runners"(Sauren 1969:238, Text 217 Rev: 19–21; D'Agostino 1997:3 Text 122, Lines 9–10); and military personnel variously termed /MAR.TU (amurrum)/ (Sigrist 1990: Text 171:6) or /àga-ús gal/ (Sigrist 1993: Text 122:5). Governors with the title of "énsi of URU×A^{ki}" are also attested. For example we can follow the career of Šulgi-zi-ñu$_{10}$, from the 44th year of the Ur III ruler Šulgi, through the eight-year reign of Amar-Sin, to the eighth year of Šu-Sin, a span of at least twenty years, ending up as the énsi of URU×A^{ki} (Steinkeller 1982:246, n. 30). Earlier, he periodically delivered tribute in animals to the imperial animal depot at Puzriš-Dagan, near Nippur. In the eighth year of Amar-Sin, he appears as an ugula or overseer of the tribute of URU×A^{ki} (Legrain 1912: Text 126, Lines 1–3). In the seventh year of Šu-Sin, he takes a mana of silver as "gun," a form of tribute required of frontier settlements, to Puzriš-Dagan as énsi of URU×A^{ki} (Genouillac 1911: Text 5515, Line 8). This individual appears as a witness together with Beli-arik, the énsi of Susa from at least the sixth year of Amar-Sin to the eighth year of Šu-Sin, on an undated text from Susa concerned with a settlement of the property of a slave (Scheil 1939: Text 242:12). He is identified merely as /^dŠulgi-zi-ñu$_{10}$. lú. URU×A^{ki}/, literally "man of URU×A^{ki}," but a term often used in late Ur III times to specify a governor. Once again this shows the close relation between URU×A^{ki} and Susa, as the former was undoubtedly within the administrative orbit of the latter at this time. Control over Susiana was lost in or after the third year of Ibbi-Sin, when Šimaški took over the area (Steinkeller 2007). This last Ur III ruler, Ibbi-Sin, claims to have launched successive campaigns late in his reign against areas east of Susa, but there is indication neither of the substance of imperial rule at Susa, such as the appointment of an énsi or maintenance of a garrison and messengers, nor of the rituals of rulership, such as restorations of or donations to temples. It seems likely these campaigns were ephemeral at best.

Local figures with Elamite names are mentioned in texts from late in the reign of Šulgi through Ibbi-Sin's reign, and some of these appear on a text from Susa written about 1800 B.C. among the twelve "Kings of Šimaški" (Stolper 1984:19–21; Gelb and Kienast 1990:317–18; Potts 1999:144–46). The third of these Šimaškian kings is Ebarat or Yabrat, who came frequently to Mesopotamia with tribute (Steinkeller 2007:218–27). However, in or just after the third year of Ibbi-Sin, Ebarat established control briefly at Susa where tablets are issued with his own year names rather than Ibbi-Sin's (De Graef 2005:99, 105–6, 112–13); thus he is no longer a tributary of the Ur III rulers. The sixth of these Šimaškian kings is Kindattu, a son of Ebarat (Steinkeller 2007:221) and probably the leader of the Elamite army that destroyed the capital of Ur and overthrew the Third Dynasty (Van Dijk 1978). There is no contemporary mention of URU×A^{ki} in the texts reporting on the final years of the ebb and flow of Ur III control in the Zagros front ranges and the rise of the Šimaški Dynasty.

In sum, when a written record of politics and economics becomes available during the mid-third millennium, the town of URU×A^{ki} is the center of a small state in a nexus of large regional states. It joins various coalitions at different times. Sometimes it is involved in war and destruction; at other times it is involved in exchange. When, however, Sargon of *Agade* and his successors impose a more enduring conquest, URU×A^{ki} becomes a minor element in the provincial structure of the empire, probably under the control of the Sargonid énsis at Susa. After the interlude at the end of the Sargonid Dynasty, the Third Dynasty of Ur reestablishes control of Susiana and (at least late in this period) URU×A^{ki} had its own governor, though relations with Susa were still close. Because of the direct relation of URU×A^{ki} with the imperial apparatus and the much greater number of published Ur III texts, more is known about this second period of imperial control. However, there is no evidence that control was qualitatively different from that of the Sargonids.

The Early to Mid-Second Millennium B.C.
(by Henry T. Wright)

Land and People in Early to Mid-Second Millennium Elam

At the beginning of the IInd millennium the types of documentation available from Mesopotamia and the surrounding regions change. Administrative texts involving rations for large groups of workers, messengers, and the like become less common, and letters, contracts, and legal texts become more common. Texts related to rulers, their rituals and their propaganda, continue. One would expect new and informative insights into early IInd millennium life in southwestern Iran. For various reasons this is not yet the case. First, the military strength of the later Šimaškian rulers, their successors the *Sukkalmaḫs*, and the successive Middle Elamite dynasties limited Mesopotamian attempts to invade southwestern Iran, so we have few accounts of military campaigns like those so informative about earlier and later times. Indeed, there is rarely any mention of the regions that are today part of Iran. Second, from southwest Iran itself we have from Susa and several nearby towns a range of letters, contracts, deeds, some administrative texts, and various royal inscriptions. There is some mention of places east of Susiana, but few mentions of places to the west such as URU×A^{ki} or *Der*. Third, unlike the IIIrd millennium texts, the large numbers of early and mid-second millennium texts are not yet available in computerized data bases facilitating the search for references to people and places.

Nonetheless, the broad outlines of the historical geography of this period are known, thanks in large part to the extensive work of François Vallat (1980, 1993). Vallat argues that from at least the late IIIrd millennium until the early Ist millennium B.C., the Elamites themselves distinguished between the lowlands of Susiana and highlands of Elam. Susa was the predominant city of the former and Anšan was the predominant city of the latter. Susa was Mesopotamian in culture and language, and though people there had both Akkadian and Elamite names, most texts were kept in Akkadian. Anšan was more Elamite in character, though the small samples of texts from the site are in various languages. This geographically and linguistically varied realm was held together by a political system in which administrators moved from local to regional posts as they became more experienced, a system that has generated much discussion (Cameron 1936; Stolper 1984).

Deh Lurān in Early to Mid-Second Millennium

Other than the brief mention of "*Arawa*," probably URU×A^{ki}, in a literary text of the ruler of Isin, *Išbi-Erra*, mentions of this toponym are rare during the IInd millennium. François Vallat (1993) has considered the issue at length, and found only two mentions, which are undated, but may be of early IInd millennium date (Scheil 1939: Texts 454:5, 424:12). Either another name had replaced the earlier name, perhaps a name actually used by its inhabitants rather than one long emplaced in the consciousness of Mesopotamian scribes, or we have been unlucky in our recovery of documents. The archaeological evidence indicates that during the Šimaški and *Sukkalmaḫḫu* cultural phases, Musiyān remained prosperous, and there were several other small towns and villages on the plain, including Baulah (DL-24) to the northwest, with the later addition of Patak (DL-35) to the southeast. During earlier Middle Elamite times, the main center on the plain shifted from Musiyān (DL-20) to nearby Tepe Gārān (DL-34), some villages were abandoned, and others such as Farukhābād (DL-32) became small hamlets. At this point, the name of the major center and the area may have changed or been transformed. But neither the local agricultural archives of *Sukkalmaḫḫu* Susa (Scheil 1913, 1939) nor the inwardly focused palace, temple, and workshop archives of nearby earlier Middle Elamite Gapnak, modern Haft Tepe (Herrero 1976; Herrero and Glassner 1990, 1991), provide useful toponymic evidence.

The Early First Millennium B.C.
(by Pierre de Miroschedji and Henry T. Wright)

Land and People in Early First Millennium Elam

During this period Mesopotamian texts refer to what is today southwestern Iran as "Elam" and its inhabitants as "Elamites." Neo-Elamite rulers term themselves "King of Anšan and Susa" (Stolper 1984:46), maintaining the ancient distinction between the highland and lowland realms. Once again there is a marked change in the nature of available documentation. While there is one fragmentary administrative archive in Elamite from Susa, the documentation in Elamite are primarily assertions about the building or restoration of temples. Most of the detailed documentation is from Babylonian and Assyrian official letters, royal annals, and other royal inscriptions. Fortunately, there is more than a century of extensive critical literature on these Mesopotamian sources.

Neo-Elamites, Assyrians, and Babylonians on the Deh Lurān Plain

During the later IInd millennium B.C. there is very little evidence of permanent settlement on the Deh Lurān Plain. Only Tepe Gārān (DL-34) has evidence of later Middle Elamite occupation and that is very limited. We have a description of the plain embedded in a text of Nebuchadnezzer I (1125–1104 B.C.), the ruler of lower Mesopotamia, granting land and privileges to LAK.TI-Marduk, a cavalry officer, perhaps an ethnic Cassite from the area of *Der*, for his loyal service in Nebuchadnezzer's second campaign in Susiana (King 1912:31–32, but following the reading of Brinkman 1968:107, n. 381 for the name); in this campaign they left *Der* in the month of Tammuz, roughly July, and marched through "intense heat scorching like flames in which horses gave out and men faltered.... The wells were without water." In about six days, they came to the *Ulai*, today the Karkheh River, and fought successfully against the Middle Elamite ruler *Ḫutelutuš-Inšušinak*. There is no mention of any towns or other places between *Der* and the *Ulai*.

When four centuries of limited documentary mention of Elam and five centuries of limited occupation of Deh Lurān ended about 800 B.C., southwest Asia was dominated by the resurgent Neo-Assyrian Empire. In Susiana, Elamite rulers resisted Assyrian attacks launched from the area of *Der*, attacks that must have come southeastward down the old route through Deh Lurān to the heart of Susiana. During this time, we have archaeological evidence of three settlements in Deh Lurān situated in a line from 'Ain Khosh (two km east of the Chikab River, and thus of the eastern limit of the Deh Lurān survey area) on the southeast to Tepe Patak (DL-35) and Tepe Gārān (DL-34) on the northeast (Miroschedji 1981b). We see the events in this area almost entirely from Babylonian and Assyrian perspectives:

- In 720 B.C., the Babylonian Chronicle reports that the army of Sargon II (721–705 B.C.) was attacked and defeated near *Der* by an army led by the Elamite ruler *Humban-Niktaš* I (Grayson 1975:73–74; Stolper 1984:45). In 710 B.C. Sargon again campaigned in the vicinity of *Der* and in western Elam, marching through nearby *Raši* and besieging the town of *Bit-Imbi*, forcing the then Elamite ruler *Šutruk-Naḫḫunte* to flee into the mountains, but did not push on to Susiana (Grayson 1975:75 ii 1–5; Stolper 1984:46; Postgate and Mattila 2004:237).

- Beginning in 704 B.C., the Assyrian ruler Sennacherib (704–681 B.C.) launched attacks against *Šutruk-Naḫḫunte*, whose successor *Hallusu-Inšušinak* campaigned in northern Babylonia (Potts 1999:269–70). In 693 B.C., however, *Hallusu-Inšušinak* was deposed and replaced by *Kudur Nahunte*, and Sennacherib raided Elam's borderlands and plundered from *Raši* to *Bit-Bunakki* (Grayson 1975:78–79; Stolper 1984:47). *Bit-Bunakki* is a district beyond *Raši*. If this district was to the southeast, it could have been Deh Lurān. However, since it was entered via a pass (Grayson 1963:91:1.34), it may have been deeper in the mountains to the northeast, perhaps the Saidmareh Valley (Young 1967:13, n. 21). When Sennacherib attacked the heart of Elam in 692 B.C., he first took twenty-six fortified settlements in *Raši*, including *Bit-Imbi*, and eleven more in the passes of *Bit-Bunakki*. Sennacherib then turned his army toward *Kudur Nahunte*'s royal city of *Madaktu* but the cold rain, snow, and swollen mountain streams stopped his advance. *Kudur Nahunte*, however, fled from *Madaktu* to *Ḫidalu*, which was protected by mountains to the east of Susiana, where he was assassinated after less than a year on the throne (Luckenbill 1924:40–44). His successor *Humban-Numena*—at the head of a large army of Babylonians, Aramaens, and peoples of various highland polities—fought Sennacherib near *Halule* on the Tigris (Luckenbill 1924:45). Accounts of the battle differ, but it is certain that afterwards Sennacherib focused his efforts on the destruction of Babylon (Potts 1999:274).

- From the later years of Sennacherib's reign, through the reign of Esarhaddon (680–669 B.C.) until early in the reign of Aššurbanipal, that is from about 688 to 666 B.C., the texts record limited diplomatic relations between Assyria and Elam.

- In 665 B.C., the Elamite Ruler *Urtak* invaded Babylonia, and in 653 Aššurbanipal (668–627 B.C.) retaliated, invading Elam from *Der* and defeating and killing the successor of *Urtak*, *Tepti-Humban-Inšušinak*, in a battle on the *Ulai* near Susa (Piepkorn 1933:59–81; Stolper 1984:50). From this campaign, there is not only an Assyrian account, but graphic representations of the campaign in reliefs with attached descriptions in Room XXXII of the Palace of Sennacherib at the Assyrian capital of Nineveh (Reade 1983:66, 99). The town of *Madaktu* is illustrated on Slab 6 of these reliefs.

It is shown as a walled town with a canal running behind it, surrounded by date palms and smaller settlements, and overlooking a river. Unfortunately, the available accounts say little of the march to the *Ulai*. After an interlude of civil war in Assyria, Aššurbanipal again attacked Elam in 647 B.C., forcing the ruler *Humban Haltaš* III to abandon *Madaktu* (Streck 1916:42ff). With the evidence of these accounts, the *Madaktu* relief, and toponymic arguments, Miroschedji (1986; 1990:60) argued for a specific location for *Madaktu* on the Deh Lurān Plain. His logic is that the place name "*Madaktu*" was transformed into "Badake," a place where the armies of the Macedonian general Seleucus camped about 312 B.C. during their campaigns against Antigonus, and this place must be near the juncture of the old road from *Der* to Susa and a road coming from the Iranian plateau to the northeast over a pass at the east end of the Kabir Kuh. If so, Badake, and therefore *Madaktu*, must have been on the Deh Lurān Plain. Furthermore, the name "Badake" was transformed into modern "Patak" (also proposed by Rawlinson 1839:91), a name given to several villages and to a mound marking a site on the eastern Deh Lurān Plain. The river shown in relief would be the Dawairij or Āb-Dānān River. The 6.5-ha mound of Tepe Patak (DL-35) had major occupations during *Sukkalmaḫḫu* and earlier Middle Elamite times, possible occupation in Neo-Elamite times, and a major occupation in Achaemenid and Post-Achaemenid times. The location of the river and canals and the occurrence of date palms would fit this location. There are two archaeological arguments against the specific identification of Tepe Patak with *Madaktu*. The first is that several examinations have produced only a few possible Neo-Elamite sherds; one would expect many. The second is that Tepe Patak is surprisingly small for a "royal city." A counter to these points is that because of time constraints, the surveyors have examined only the citadel and the site may be richer and bigger than it now seems. Certainly the toponymic arguments imply that *Madaktu* was somewhere in this area.

- The following year Assyrian armies returned to finish the destruction of Elam. Once again they marched through *Der*, and once again *Humban Haltaš* fled eastward from *Madaktu*. The Elamite forces took positions on the Susiana Plain at the *Idideh* or *Hudhud* River, probably the modern Dez River (Waters 2000:75–76) near the Middle Elamite center of Dur-Untash (Choghā Zanbil), little occupied but still ritually important. However, they abandoned these positions and fled into the mountains of *Ḫidalu*. The Assyrian troops were able to take towns and captives all the way to the borders of *Ḫidalu*. They turned back westward and on the return, for the first time in their many campaigns, thoroughly sacked and burned Susa (Streck 1916:46ff; Stolper 1984:51; Potts 1999:284–85). In spite of this, a letter from the Assyrian general *Bel-Ibni* to Nineveh reports that *Humban Haltaš* returned to take the throne at *Madaktu*, then fled again southward across the *Ulai* River (Waters 2000:78). After these events, Elam rarely appears in the records of Babylonia and Assyria.

There is, however, scattered evidence showing that some form of Elamite polity had survived. Texts from Susa attest to the administration of craftspeople and relations with towns to the east (Scheil 1907; Vallat 1993). Survey has located a number of settlements thought to be later Neo-Elamite on the Susiana Plain (Miroschedji 1990:59) and in the valleys to the east (Carter 1994). Also, members of important families were interred with great wealth in *Ḫidalu* to the east (Alizadeh 1985). There is, however, little evidence of the final phase of the Neo-Elamite period in the northwest, including the Deh Lurān Plain. A fragmentary text suggests that Elamite troops were in action in northern Babylonia during the reign of Nebuchadnezzar II (605–552 B.C.) and he and his successors may have briefly controlled Susa (Stolper 1984:54; Potts 1999:290–94). If so, Elamite and Neo-Babylonian armies and administrators must have taken the common route between *Der* and Susiana crossing the Deh Lurān Plain.

The Mid-First Millennium B.C.
(by Henry T. Wright)

Land and People In Mid-First Millennium Elam

Elam was a tiny part of the realm of the Achaemenids, an empire that extended from the Indus Valley to the oases of the Libyan Desert. In spite of the importance of this empire in world history, few Achaemenid archaeological sites are well studied and rich contemporary archives of the Achaemenid administration are available primarily from the summer capital of Persepolis, from the important provincial capital of Babylon, and from Egypt. We are dependent on the later redactions of views of other peoples—some within the Empire as in the case of the Israelites, and some without as in the case of many of the classical Greek writers. The Achaemenid geography of what is today southwestern Iran is known from both Old Persian and Greek sources. Susiana was organized as the satrapy or province of Khuja (Herzfeld 1968); the Khujian people were perhaps descendants of the Elamites and perhaps later recognized by the Greeks as the Uxians (Potts 1999:349–51). This area and the mountains to the northeast were occupied by Elamite speakers, who several times rebelled against Achaemenid rule, but who also provided tribute, troops, and scribes who kept Achaemenid government records in the Elamite language throughout much of the eastern empire. The land northwest of Susa—modern Luristan and Pusht-i Kuh including Deh Lurān—was Cissia, the land of the Cossai, thought by some scholars to be the descendants of the Cassites who dominated Mesopotamia during the mid-second millennium B.C. (Herzfeld 1968). A rugged land of oak forested mountains and high pastures, this area seems to have had little interest for Achaemenid authorities.

Deh Lurān under the Achaemenid Dynasty

If Elam was a small part of the Achaemenid realm, Deh Lurān was insignificant. A line of settlements—evenly spaced from southeast to northeast across the plain about 17 km apart—were no doubt stations on the "Royal Road." One of these, Tepe Gārān (DL-34), was a major town of 17 ha with indications of Achaemenid public architecture. However, without contemporary archives from a nearby center such as Susa, we must rely on the indirect observations of Greek writers.

Herodotus reports seeing a map of the Achaemenid Empire on a bronze plate in the possession of Hecateus of Miletus. He may have derived his account of the Royal Road from Sardis to Susa in part from this source, but also from his own observations. Whatever the sources, he says that after the Matienien country, which appears to be the Diyala Region of Babylonia, "you find yourself in Cissia, where eleven stations and forty-two and one half parsangs bring one to another navigable stream, the Choaspes, on the banks of which Susa is built" (Herodotus V:52). Herzfeld (1968) argues that the border of Cissia was southwest of *Der*. The distance by the modern road from Mehran, just southwest of *Der* and inside the Jebel Hamrin, to the Karkheh River is 205 km, and thus the stations would have averaged 17 km apart, in close agreement with the observed spacing in areas we have surveyed. Unfortunately, these stations are neither named nor described.

There is another mention in Herodotus' "Persian Wars" that may refer more explicitly to the Deh Lurān area. In discussing the fate of defeated Greek peoples, Herodotus (VI:119) notes that the citizens of Eretria, a city on the island of Euboea in the Western Aegean, were taken to Susa, where Darius settled them in "one of his own stations in Cissia, a place called Ardericca, 210 stades [approximately 40 km] from Susa, and 40 stades [approximately 8 km] from a well that yields a produce of three different kinds. From this well they get bitumen, salt and oil." Herodotus goes on to describe how these three materials are separated and prepared. This place called "Ardericca" is not the same as a village said to be on the Euphrates in Assyria (Herodotus I:185). There are two common interpretations of these geographical observations. First is that the distance given is correct and Ardericca is about 40 km from Susa toward Cissia, that is to the northwest. The problem is that there is no source producing bitumen in this location. Rawlinson (1839:93) does report visiting a small bitumen source near Kir-Ab, about 25 miles north of modern Dezful, or 60 km north of Susa. The area has not been surveyed recently and we do not know if there are Achaemenid sites there. Second is that the distance given by Herodotus is wildly off and that Ardericca was near a known oil and bitumen source in the upper Diyala Valley (Herzfeld 1968). In fact, there is a major source of bitumen, oil, and salt in Deh Lurān at 'Ain Qir or Chešme Gir, as discussed above in our consideration of the IIIrd millennium documentary record. The modern source is 7 km north of the large Achaemenid settlement of Tepe Gārān (DL-35), which is approximately 105 km or 550 stades by road from Susa, and there are traces of ancient sources up to 12 km away, bracketing the 8 km indicated in the text of Herodotus. It is possible that Tepe Gārān was Achaemenid Ardericca. If so, we know from Herodotus that the people there spoke Greek at the time of his travels in the mid-fifth century B.C., but that by the time of Alexander's conquest about 325 B.C., Quintus Curtius Rufus states that they spoke both the local language and a degenerate form of Greek (Briant 2002:722). It is possible that Ardericca may have been a town referred to in Pre-Achaemenid sources as "*Urdalika*," which Aššurbanipal's annalist includes in an unordered list of Elamite towns destroyed in 647 B.C. (Streck 1916:42). This is perhaps the same place mentioned in a Neo-Elamite economic text from Susa, reconstructed as /U[r-k]a-da-a-r[a-na]/ by Scheil (1907:146: Text 164:7) Unfortunately, these mentions offer little additional information, and we have found no other mentions of Ardericca or *Urdalika* in published sources.

Chapter 8

A Land Between

Reflections on Early Historic Deh Lurān

Henry T. Wright and James A. Neely

A small, poor foothill valley, the Deh Lurān Plain was a place of little value to the dynasts, soldiers, messengers, traders, and other travelers who passed through it from the late IIIrd to early Ist millennia B.C. It was a plain of no significance between more important places such as Susiana and Babylonia. When larger regional and trans-regional polities developed in these and other richer heartlands, Deh Lurān flourished. When these polities disintegrated, the plain was sometimes largely or completely abandoned. Nonetheless, the early historic inhabitants on the plain gradually transformed their way of life. Let us begin this final note by summarizing the framework of successive political entities within or between which these inhabitants tried to make a living. We will then turn to some specifics of production and communication in the Deh Lurān area.

Deh Lurān in the Trans-Regional Context

During the first seven centuries of the IIIrd millennium, Deh Lurān was the center of a small independent polity centered at the town marked by Tepe Musiyān (DL-20), perhaps ancient URU×A^{ki}. Its local environment was relatively rich, with an easily maintained canal system and access to a number of widely valued commodities such as flint, limestone, sandstone, bitumen, and mountain hardwoods. During the conflicted times of the later Early Dynastic period, a new fortified town on the east end of the plain (DL-41) was built and briefly inhabited. Later in the IIIrd millennium, the plain was absorbed into trans-regional polities, first by the Sargonid Dynasty of Agade and subsequently by the Third Dynasty of Ur. The Šimaški ceramics used in settlements of the Deh Lurān and Susiana Plains are similar to Late Akkadian and Ur III ceramics in Mesopotamia, and the textual evidence indicates URU×A^{ki} was attached to the provincial administration in Susa until the last decades of the Ur III Dynasty when it had its own governor.

During the IInd millennium, the plain fell under the control of at least five successive Elamite dynasties, which incorporated regions in both the high Zagros and the lowland foothill valleys. All these polities seem to have been strong enough to keep Mesopotamian armies at bay, and we lack Mesopotamian sources on the area. Unfortunately, we have not yet found the Elamites' own textual records of their western frontiers in the Deh Lurān area. Though they claimed rule over several very different regions, as far as we know, these Elamite rulers did not make claims to universal suzerainty, did not promulgate law codes, and did not deify themselves while living. At least during the earlier IInd millennium, under Šimaški and Sukkalmaḫḫu rulers, Deh Lurān was relatively prosperous. During the latter phase its population rose to perhaps 8000 people, greater than during Early Dynastic times, and its consumer goods were identical with those of Susiana. The large town of Musiyān was probably fortified and had major central buildings. Subsidiary towns on the east (DL-35) and west (DL-24) flourished. However, there were few smaller settlements, perhaps a consequence of the plain's position near a potentially unstable frontier. During the later IInd millennium, Middle Elamite dynasts continued to make claims for trans-regional rule and they built several new capitals, but we

113

know little of their social and economic organization. On their western frontier, there are indications of retrenchment. Musiyān and most other settlements were abandoned, and a smaller center developed to its northwest at Gārān (DL-34). Even this center may not have been inhabited toward the end of the millennium during later Middle Elamite times.

At the beginning of the Ist millennium B.C., there is no evidence of occupation on the Deh Lurān Plain, but subsequently Neo-Elamite rulers reestablished several settlements in a linear arrangement from southeast to northwest along the edge of the alluvial slopes. Their locations are not in former centers of irrigation and cultivation, and it is likely that they were maintained to service a route of passage. One of the fortified "royal cities" of the Neo-Elamite kings, Madaktu, destroyed during Assyrian incursions, may have been on the east end of the Deh Lurān Plain. With the creation of the Achaemenid Empire by Cyrus, Deh Lurān became a tiny part of a polity extending from the Levant and Anatolia on the west to Afghanistan on the east. Cyrus successors incorporated Egypt on the west and much of the Indus Valley on the east. The Achaemenid dynasts certainly sought universal sovereignty over the civilized world as they knew it, controlled a network of core territories with a complex administrative and military apparatus, connected them with a network of routes and messenger stations, and promoted one of the world's first universal religions. Later, perhaps after the reorganization of the empire under Darius and Xerxes, the Neo-Elamite settlements on the Deh Lurān Plain were reestablished, and Gārān became a large town, probably with major public buildings. These Achaemenid settlements—evenly spaced at the regular 17-km interval reported by Herodotus for this part of the Royal Road from Susa to Sardis (as discussed in Chapter 7)—were most certainly intended as service points for travelers. If Gārān was the "Ardericca" of the account of Herodotus, Deh Lurān profited from the resettlement of captive peoples near the imperial heartland, a policy found in early empires throughout the world. After the campaigns of Alexander, the infrastructures created by the Achaemenids passed under the control of the Seleucids, as we will discuss in our next volume.

In sum, if we view empires as large trans-regional polities whose elites dominated their world as they knew it, built systems of communication, bureaucratic control, and military force, and promulgated ideologies of universal sovereignty, law, and even cosmology, then during at least three of the periods covered in this monograph, the Deh Lurān Plain was an integral part of an early empire, and during several other periods it was dominated by large regional polities whose full span of control remains ill-defined. We now turn to issues of resource appropriation and communication.

Transforming Water Management and Food Production on the Deh Lurān Plain

By the early IInd millennium, if not before, the plain had become a much poorer place than it had been in the time of village societies and of the first states. Several perhaps interrelated processes caused this to happen. First, the climate was generally more arid than during the Vth to IIIrd millennia, at times probably drier than the semi-arid conditions with predominantly winter rainfall of today. Second, the rivers, which had aggraded—moving back and forth across the plain depositing silts and gravels—cut down into these deposits, entrenching their floodplains 4–6 m below plain level and lowering the water table (Kirkby 1977). This would have made it more difficult for people to get the waters of the Mehmeh and Dawairij (Āb-Dānān) Rivers to their fields. They would have had three possible solutions. The first would be to cut the canal offtakes deeper into the alluvial fans where the rivers leave the foothills and enter the plain. The second would be to move the offtakes farther upstream where the rivers were not so deeply cut, thereby necessitating the excavation of longer canals. The third would be to dig tunnels through the fan to feed the canals on the plain. These tunnels—known in Persian as "*qanāt*"—were not the *qanāts* of the sort well known throughout the Iranian plateau (English 1966; Sajjadi 1982), as they tap rivers rather than the water table at the foot of the mountains. However, the use of tunnels with periodic vertical well shafts or vents used in construction and subsequent maintenance to tap rivers is similar and may even have been an experimental step in the development of classical *qanāt* technology. Some of the Deh Lurān examples may be even closer to true *qanāts* given Neely's (1974b) suggestion that the upper reaches of some tunnels run parallel to the river for some distance, perhaps conserving an intervening filtering barrier of soils and gravel or conglomerate so that the heavy sediment load of the Mehmeh and Dawairij Rivers would not enter the tunnels. The *qanāt* headwork feeding the canal to Patak (DL-317) is such a filtering *qanāt*, and probably dates to Neo-Elamite times. There is at least indirect evidence that tunnel technology was already in use earlier in the IInd millennium B.C. Tepe Gārān (DL-34) was already fairly large in Middle Elamite times, and the tunnel (DL-121) bringing water from the Dawairij (Āb-Dānān) may have been necessary both for domestic uses and to irrigate nearby fields and gardens. Unfortunately, erosion prevents us from knowing whether or not the DL-121 *qanāt* was of the filtering type. It seems even more unlikely that the large Achaemenid town at Tepe Gārān could have been supplied with water and food without the use of *qanāts*. The resolution of the earliest date of *qanāt* use on the Deh Lurān Plain will require excavation and direct dating of the deposits in and the sediments around well shafts and on the canal banks.

If the Deh Lurānis had learned to get water to their fields, what did they grow? To the traditional Near Eastern crops long grown—for example wheat, barley, lentils, chickpeas, grapes, and dates—a diversity of new crops, mostly from South Asia (for example, rice, sugar cane, and citrus fruits), were added sometime during the IInd or Ist millennium B.C. One could argue that the limited summer water supplies would make it impossible to grow useful quantities of rice and sugar cane in Deh Lurān, and that the most likely crops would have been wheat and barley with limited additions of fruits such as citruses or grapes. However, few Elamite sediment samples and no later samples were processed in Deh Lurān with the flotation method, and very few seeds have been identified. Clearly, future excavators of early historic sites must recover flotation samples for botanical study, following the example of previous excavators' work on pre- and proto-historic sites.

Reorienting the Deh Lurān Settlements around Corridors of Communication

The Deh Lurān Plain was on the transport routes between the valleys of the Zagros front ranges and the Lower Mesopotamia heartland at least since the late Vth millennium, but there is no clear indication of site alignments or hollow ways that would mark earlier routes. One could argue that Musiyān (DL-20) dominated a crossing of the Dawairij (or Āb-Dānan) and that Farukhābād (DL-32) dominated a crossing of the Mehmeh, but there is no indication other than location for these assertions. When the main line of transport, as indicated by a linear arrangement of sites, shifted to the fringes of the piedmont on the northeastern edge of the plain (as discussed above), the visibility of such features in the available imagery improves. Some of the linear features articulate with *qanāts* and must be canals. Others could be roads. However, we do not have the possibility of reexamining them on the ground. We have shown the possible Achaemenid route on Figures 5.6 and 6.6, and noted that this may follow the line of an earlier route of Neo-Elamite times. As we will discuss in our final Deh Lurān volume, this route continues to be important in later periods, and indeed, is approximately the location of the main road today. Another future task for archaeologists working on the Deh Lurān Plain will be the excavation of cross-sections across possible roads within and between settlements to precisely define them and directly date them based on geological evidence and ceramic associations.

The specific locations of the roads that constituted routes in any particular period are perhaps less interesting than the question of how settlements placed in the locations dictated by the needs of travelers were actually sustained. Food and water for people may have been minimal, but the water and fodder necessary for substantial numbers of pack animals would have required increased canal capacity and productive fields. Likewise, the water and food required for the movement of armies would have been a burden, but in these rare events civilian populations probably fled to the mountains. Considering the settlements thought to be on the Achaemenid route from southeast to northwest, we can get an idea of the accommodations made to provision travelers and their animals as they passed through the Deh Lurān area.

- 'Ain Khosh, just east of our survey area, has springs with, as the name implies, "sweet" water. It is also near the Chikab River, a smaller tributary of the Dawairij (Āb-Dānan), which has fresh water much of the year. The springs create an extensive meadow on which animals could graze.

- Tepe Patak (DL-35) has no natural water supply and no natural summer pasture. It is 3.5 km north of the Dawairij (Āb-Dānan), to which animals would have to be driven if there was no canal. Isolated Patak was sustained by a 16.8-km long canal (DL 317), flowing eastward from the left bank of the Dawairij. That canal had originally been excavated to support several Early Dynastic sites, but was refurbished as early as Sukkalmaḫḫu times to support Patak alone. Perhaps as early as Middle Elamite times, and certainly by Neo-Elamite times, the headwork was a *qanāt*. Present evidence indicates Patak was rather small and it seems unlikely that its inhabitants could have completed the annual cleaning and maintenance of the canal (as well as its many smaller field canals not detectable on our images) by themselves. This would have required that administrators provide extra labor. In later times, administrators had to have provided the specialists who have traditionally overseen the construction, cleaning, and repair of *qanāts* (English 1966).

- Gārān (DL-34) could have had access to the water of the Ab-i Garm, near modern Deh Lurān to its northwest, and 'Ain Qir or Chešme Gir, to its north, both mineralized and foul tasting, but potable and certainly adequate for animals. Even during the dry summer, pasture would have been available in meadows around these water sources, as well as along the Dawairij, 4.5 km to the east. The most accessible source of sweet water was via a relatively short 6.5-km canal (DL-121) from the right bank of the Dawairij River (Āb-Dānan). As with the source of water for Patak, perhaps as early as Middle Elamite times, and certainly by Neo-Elamite times, the headwork for this canal was a *qanāt*. The population of Gārān, at least by Achaemenid times, was great enough to maintain this relatively short canal, though specialists probably would have been needed to clean and repair the *qanāt*.

- Tāfuleh (DL-8), an occupation newly founded in Achaemenid times, was substantially to the west of the meadows of the Ab-i Garm surrounding the long abandoned site of Bauleh (DL-24). As discussed in Chapter 6, the only practical source of water for Tāfuleh was the Ab-i Garm, and a canal (DL-331) was extended 6 km to Tāfuleh, the requisite distance of 17 km from Gārān. The estimated population of Tāfuleh could have been enough to maintain this canal, and no *qanāt* requiring specialists was involved. The major problems would have been the poor quality of its water and the limited pasture that could be created by flooding low areas southwest of Tāfuleh. Animals could have been driven 6 km to the Mehmeh River, but this river has limited and mineralized water later in the summer and autumn and may have often been inadequate for the needs of travelers.

- The next stage of the road would cross the Mehmeh, and could have gone by several possible paths. We can see no evidence of permanent water sources or canals on any of these possibilities on the available images. Reexamining Tāfuleh and tracing the Achaemenid and earlier routes westward from Tāfuleh are obvious priorities for future field workers.

The settlement evidence is inadequate to even suggest how the Neo-Elamite route was administered and maintained (if it was maintained at all), but it seems likely that the Deh Lurān portion of the Achaemenid road was administered from the large central site of Gārān, and that extra labor could be recruited and sent to clean canals or otherwise repair the road where needed. Whether *qanāt* specialists lived at Gārān or were requested from settlements farther east is not known. Investment in routes for the passage of officials and goods is common in developed imperial formations, and the maintenance of routes and facilities in Achaemenid times is not surprising. Whether such systems will be demonstrated for earlier imperial formations remains to be seen.

The Implications of this Survey for Future Archaeological Work

Throughout the preceding chapters we have made a number of suggestions regarding observations we should have made or might now be made with new methods. It is appropriate to end these reflections with a brief summary of these suggestions.

1. The vicissitudes of motor vehicles, health, and so on prevented us from surveying the far western extremity to the west of the Mehmeh River, and the examination of the southeastern area of the plain was hasty. Though agricultural land leveling will have made survey more difficult, proper examination of these areas is an important goal.

2. Several key sites lack certain observations that could be quickly made in the course of brief re-survey. For example, Early Dynastic DL-41 lacks sector collections so we cannot be certain of its occupational area. Also, the rectangular group of possible public buildings in its center (visible on the 1961 imagery in Plate 6) should be checked. In addition, the area around Patak (DL-35) requires reexamination in order to see whether or not the settlement was ever larger than the mounded area. The area north of Gārān (DL-34) similarly requires examination, as Achaemenid sherds were found on small low mounds in this area. Gārān needs reexamination in late March when the first grass is coming up, and a plan could be made of large gravel-founded buildings seen at that time in 1968. Finally, Tāfuleh requires a sectorial collection to differentiate the Achaemenid from the Seleucid areas.

3. Our survey was conducted before methods for recognizing pastoral camps were developed. Future intensive survey on the piedmont slopes should look for possible campsites.

4. Both Musiyān and Gārān are prime candidates for more intensive research. Excavation would resolve basic inadequacies in our ceramic sequences, provide direct evidence of crops grown and animals herded, document the later history of the use of Deh Lurān's bitumen source, and might well produce cuneiform records that would resolve the toponymic and historical issues raised in Chapter 7.

5. A program of geological cuts across possible canals, *qanāt* shafts, and roads would have many uses. Carefully recorded sections of canals would indicate their slope and maximum possible volume. Pottery samples from their fill would date their final period of use. Sediment samples from debris thrown from them could be used for Optically Stimulated Luminescence (OSL) dating to determine when they were built (Aitken 1998). The radiocarbon dating of organics trapped in the travertines and other minerals coating the walls of canals and *qanāts* could further confirm their dating (Caran et al. 1995; Winsborough et al. 1996). Carefully recorded sections of roads would show their width and evidence of repair, if any, and OSL dating of road surfaces might be possible as well.

With an outstanding generation of young Iranian archaeologists now working throughout the country, we have every hope that some of these goals will be accomplished sufficiently soon to surprise us with their new discoveries and insights.

Appendix A

Counts of Artifacts in Survey Collections

Table A1. Musiyān (DL-20) Jemdet Nasr ED I-ED III ceramics.

Types		1	2W	2E	3-1	3-2	3-3	4-1,2	4-3	5-1	5-2	5-3	6E	6C	6W
Sandy Ware															
ED-RC-1	conical cup rim	7	2	1	1	4	-	-	1	-	1	1	2	-	8
ED-RC-2	incurved cup rim	-	-	-	-	1	-	-	-	-	-	-	1	1	-
ED-BC-1	solid footed cup	8	3	6	3	4	7	8	7	5	6	-	29	3	15
ED-BC-2	narrow cup base	6	2	-	5	5	9	-	1	3	3	-	1	-	1
ED-BC-3	wide cup base	4	1	3	10	15	6	-	-	4	2	-	-	1	1
ED-RB-1	round lip bowl	1	-	-	1 Tk	1	-	-	-	-	-	-	1 Red	-	-
ED-RB-2	beveled lip bowl	3	1	1	1	1	-	-	-	1	-	-	-	-	1
ED-RB-3	ledge rim bowl	1	-	1	1, 1 Gr	4	1 Red	-	1	1	2	1	1	1	-
	other bowl	-	-	-	-	-	-	-	-	-	-	-	-	-	-
ED-RJ-1	round lip jar	2	-	6	3	2	-	1	-	3	1	3	1, 1 Red	-	1
ED-RJ-2	thickened round lip jar	1	-	1	1	-	-	-	-	1	1	-	2	-	-
ED-RJ-3	flared expanded jar	2	3	1	-	-	-	-	2	1	3	-	-	-	2
ED-RJ-4	ledge rim jar	4	2	2	3	3	-	-	1 Red	2	-	1	2, 2 Red	1	3
ED-RJ-5	band rim jar	5	1	1	1	1	-	-	-	2	-	2	1	-	-
ED-RJ-6	small carinated pot	-	-	-	-	1	-	-	-	1	-	-	-	-	-
ED-O-1	plain strip	5	2	1	6	1	2	2	1	5	1	-	2, 1 Red	12	6
ED-O-2	hatched strip	15	4	7	10	5	1	-	-	4	-	-	15	8	5
ED-O-3	monochrome lines	-	-	-	-	-	-	1	-	2	-	-	-	-	-
ED-O-4	polychrome lines	1	-	-	-	-	-	-	-	-	-	-	2	-	-
ED-O-5	monochrome motif	-	-	-	1 Gob Sh	3 Gob Sh	-	4, 1 Gob Sh	1 Gob Sh	-	-	-	4, 2 Gob Sh	-	1
ED-O-6	polychrome motif	-	-	-	-	-	-	1	-	-	-	-	2	-	-
ED-O-7	conical spout	2	-	-	-	3	-	1	-	2	-	-	4	1	-
ED-O-8	lug	-	1 Cy, Sl	1	-	-	-	-	-	-	-	-	2 Vr, Hp	-	1 Hz, Hp
ED-O-9	fenestrated stand	-	-	-	1 Hz, Vp	1	-	1	1	1	1	-	1	-	1?
ED-B-2	flat base	3	1	-	1	-	4	-	-	1	1	-	-	-	-
ED-B-3	turned ring base	1	1	1	-	-	1	-	-	-	-	-	-	-	-
ED-B-4	high turned ring base	6	3	2	4	6	5, 1 Red	-	-	4	1	-	1 Red	-	-
ED-B-5	pinched ring base	4	1	2	2	5	3	-	-	3	2	-	-	-	-
ED-B-6	fine ring base	-	-	2	1	1	-	-	1	1	-	-	2 Red	-	-
Gray Ware															
	jar rim	2	2	2	-	-	-	-	1	1	3	1	1	1 Red	2
	body	-	1	-	-	-	-	-	-	-	3, 1 Shell	1	1	-	1
Other				2 Discs 2 Fgrns							1 Strainer 1 Ring Scraper				

Key: Tk: thick; Gob Sh: Motif on goblet shoulder; Gr: grooved; Frgns: figurines; Red: red slip
Lugs: Cy: cylindrical; Hz: horizontal; Hp: horizontal perforated; Vr: vertical; Vp: vertical perforated; Sl: solid; Pr: perforation

Table A2. Musiyān (DL–20) Elamite ceramics.

Types									Areas							
		1	2W	2E	3-1	3-2	3-3	4-1,2	4-3	5-1	5-2	5-3	6E	6C	6W
Sandy Ware																
SI-RC-3	indented rim cup	3	-	1	1	3	-	-	-	6	-	2	5	1	2
SI-RB-1	round lip bowl	3	-	-	-	-	-	-	1	-	-	-	-	-	-
SI-RB-3	ledge rim bowl	-	1, 1 Wvy Ln	2	-	1	-	-	-	1	3, 1 Wvy Ln	1	1, 3 Wvy Cm	-	-
	carinated bowl	-	-	-	3	-	-	-	-	-	-	-	-	-	-
SI-RJ-1	round rim jar	-	-	-	3, 1 Tk	-	-	-	-	-	-	-	1	-	3
SI-RJ-2	oblique rim jar	-	1 St Cm	-	-	1	-	-	-	1	-	1	-	-	1
SI-RJ-3	expanded band jar	-	1	-	3	7	-	-	3	3	14	1	1	-	1
SI-RJ-4	band rim jar	3	-	-	1	2	-	-	-	2	-	2	1	-	-
SI-RJ-6	ledge rim jar	3	-	2	3	2	1	-	-	1	3	1	4	-	2
SI-RJ-7	grooved rim jar	3	1 Wvy Cm	3	4	4	-	-	-	5	9	-	2	-	-
SI-RB-3	ledge rim basin	5	2	3 Cm	4	3	6	-	2	3	-	2	3	6	1
	other basin	-	1 Exp Cm	-	-	1 Sq	-	-	-	-	-	-	-	-	1 Sq
SI-B-1	flat/stump base	4	3	5	3	9	3	-	-	8	-	7	1	-	-
SI-BG-1	button base	1	3	1	-	-	-	-	-	-	-	4	-	-	-
	ridged shoulder	2	4	2	5	-	1	-	1	1	-	-	1	-	2
SI-O-2	combed body	4	1	3	1	6	3	-	-	1	2	-	5	1	-
	thumb-impressed strip	-	1	5	4	4	2	-	-	5	1	-	11	1	2
	other					5 Whorls	3 Trays			3 Strainers					
Fine Ware																
SM-RG-1	goblet rim/shoulder	-	-	-	4	1	5	-	1	1	7	-	-	2	-
SM-BG-1	button base: plain	-	-	2	-	2	4	-	-	6	8, 1 Sm	-	-	-	-
SM-BG-2	button base: ridged	-	-	-	1	-	-	-	-	1	2	-	-	-	-
Vegetal Ware																
SM-RB-3	flat rim bowl	-	-	-	-	-	-	-	-	2	1	-	2	1	-
SM-RJ-1	ledge rim jar	2	-	2	1, 1 Lg	-	-	2 Lg	-	4	8	-	2	1, 1 Lg	1
SM-RJ-5	band rim jar	-	2	-	2	-	-	-	-	3	1	-	2	-	3
SM-RJ-2	square rim jar	-	-	-	-	-	-	-	-	-	3	-	2	1	-
SM-RB-5	ledge rim basin	2	5	7, 1 Cm	5	-	2	4	1	4	16	-	10	5	5
SM-RB-6	square rim basin	-	1 Sm	2	2	-	4	1	-	1	2	-	6	-	-
SM-RB-7	band rim basin	-	1	-	-	-	-	-	-	-	-	-	1	-	-
SM-B-1	flat/stump base	2	5	1	3	-	8	1	-	9	-	14	-	-	-
SM-B-2	low ring base	1	1	1	1	-	-	-	-	-	-	-	-	-	-
SM-B-3	high ring base	1	1	2	2	-	2	-	-	2	-	-	-	-	-
Other		1 Wvy Cm 1 St Cm	2 Wvy Cm	-	-	-	-	Fgrm	-	-	2 Cylindrical Drains	-	4 Wvy Inc 1 St Cm	1 Wvy Cm 8 Discs	1 Wavy Cm

Key: Tk: thick; Exp: expanded rim; Sq: square rim; Sm: small; Lg: large; Fgrm: figurine
Wvy Ln: wavy line; Wvy Cm: wavy combed; Wvy Inc: wavy incised; St Cm: straight combed; Cm: combed; Exp Cm: expanded, combed

Table A3. Tenel Ramon (DL-27) Elamite ceramics.

Types		1-1	1-2	1-3	1-4	2-1	2-4
				Areas			
Sandy Ware							
SI-RB-3	indented rim cup	-	-	-	-	-	1
SI-RJ-1	round rim jar	-	-	-	-	1	-
SI-RJ-2	ledge rim jar	1	-	1	-	2	-
SI-RJ-3	band rim jar	-	-	-	-	-	-
SI-RJ-6	grooved rim jar	-	-	-	1 Lg	-	-
SI-B-1	flat base	3	-	-	-	2	-
SI-B-2	ring base	-	-	-	-	1	-
Vegetal Ware							
SM-RB-1	plain bowl	1	-	-	1	1	-
SM-RG-1	goblet rim/shoulder	1	-	1	5	4	-
SM-BG-3	button base: small	-	-	-	-	2	1
SM-BG-4	button base: plain	2	1	8	1	8	2
SM-BG-2	button base: ridge	-	-	-	2	1	-
SM-RJ-1	round rim jar	-	1	-	-	1	1
SM-RJ-3	ledge rim jar	3	-	3	2	7	1
SM-RJ-4, 5	band rim jar	1			5	-	1
SM-RJ-2	square rim jar	-	-	-	3	1	2
SM-RB-5	flat rim basin	1	3	2	6	2	-
SM-RB-6	square rim basin	-	-	2	3	3	2
SM-RB-7	band rim basin	-	1	-	-	1	3
SM-RB-4	grooved rim basin	-	-	-	1	-	-
SM-B-7	flat stump base	4	2	5	5	-	-
SM-B-2	low ring base	-	2	-	3	1	2
SM-B-3	high ring base	1	1	1	-	-	-
SM-O-3	combed body	2	2	-	-	2	-
SM-O-4	grooved body	1	6	1	3	6	-

Key: Lg: large

Table A4. Tepe Gārān (DL-34) Elamite and Achaemenid ceramics.

Types		1X	1¢	2	3	4	5	6	0
Elamite Ceramics: Vegetal Ware									
SM-RB-7	band rim basin	-	2	-	-	1	-	-	1
SM-RB-5	ledge rim basin	-	-	-	1	1	2	1	-
SM-RJ-5	band rim jar	-	-	1	-	-	-	-	-
SM-RJ-5	heavy band rim jar	1	-	1	-	1	-	-	1
SM-RJ-3	oblique rim jar	-	2	-	-	2	-	-	-
SM-RJ-2	square rim jar	1	-	1	-	-	-	-	-
SI-RJ-7	grooved rim jar	-	1	-	-	-	-	-	-
SM-RB-1	round lip bowl	-	1	-	-	1	-	-	-
SM-RB-5	carinated bowl	-	1	-	-	-	-	1	-
Achaemenid Ceramics: Hard Sandy Ware									
AM-RJ-3	beaded rim jar	2	-	-	1	2	2	1	5
AM-RJ-5	beaded rim jar, ridge	1	1	-	-	-	1	-	-
AM-RB-1	beaded rim bowl	-	-	2	1	-	1	-	2
AM-RB-2	carinated beaded rim bowl	1	-	-	1	1	1	2	1
AM-RB-4	carinated ledge rim bowl	-	1	-	1	-	1		
AM-RC-2	indented rim cup	-	-	2	-	-	-		
Other									
		2 trays	1 plate		1 plate				

Table A5. Tepe Patak (DL-35) Elamite and Achaemenid ceramics.

Types		1	2	3	4	5	6	7	9	10	11	12	13	14	15	16	17
Elamite Ceramics: Sandy Ware																	
	goblet rim	-	-	-	-	-	-	1	-	-	-	-	-	-	-	-	-
	plain button base	-	-	-	-	-	-	-	-	-	-	-	-	-	1	1	-
SI-RJ-1	round lip jar	-	-	1	-	-	-	-	-	-	-	-	-	-	-	-	-
SI-RJ-3	band rim jar	-	-	1	-	-	-	-	-	-	-	-	-	-	-	-	-
	ledge rim basin	-	-	-	-	-	-	-	-	-	-	-	-	-	-	2	-
SI-B-1	flat base	1	-	1	-	-	-	1	-	-	-	-	1	-	-	-	-
SI-B-2	ring base	-	-	-	-	1	-	-	-	-	-	-	-	-	-	-	-
Elamite Ceramics: Vegetal Ware																	
	flat rim bowl	-	1	1 Grv	-	-	2	-	-	-	-	-	1	-	-	1	-
SM-RG-1	goblet rim/shoulder	-	1	-	-	1	-	-	-	2	-	3	10	1	-	2	-
SM-BG-3	button base: small	-	-	-	-	-	1	-	-	-	1	8	4	-	-	1	-
SM-BG-4	button base: plain	-	1	-	2	1	3	-	1	3	4	24	7	-	1	1	1
SM-BG-2	button base: ridge	1	2	-	-	-	-	-	-	2	-	3	1	-	-	-	-
SM-RJ-3	ledge rim jar	-	3	-	-	-	-	-	-	-	1	4	5	1	-	-	-
SM-RJ-4,5	band rim jar	-	3	3	3	1	6	1	-	1	-	9	9	3	-	3	2
SM-RJ-2	square rim jar	1	2	-	-	-	-	-	-	-	-	-	1 Inc	-	-	-	-
SM-RB-5	flat rim basin	1	1	-	1	1	1	-	-	-	3	2	2	-	-	-	-
SM-RB-6	square rim basin	1?	-	-	-	1	1	1	-	-	1	1	-	1	1	-	-
SM-RB-7	band rim basin	-	2 Grv	-	2	-	-	1	-	-	2	-	2	-	-	-	-
SM-B-1	flat/stump base	-	1	-	-	-	1	-	-	-	-	2	5	1	-	-	-
SM-B-3	high ring base	1	-	-	-	-	2	-	-	-	-	-	1	-	-	-	-
SM-O-3	combed body	-	1	-	-	-	-	-	-	-	-	2	-	-	-	-	-
SM-O-4	grooved body	-	2	-	-	-	3	-	-	-	-	1	-	-	-	-	-
SM-O-2	incised body	-	-	-	-	-	-	-	-	3	1	-	-	1	1	-	-
	plain ridge	3	-	-	-	-	-	-	-	-	-	-	-	-	-	-	-
Achaemenid Ceramics: Hard Sandy Ware																	
AM-RJ-3	beaded rim jar	-	-	-	-	-	-	-	-	-	-	-	1	1	-	1	-
AM-RJ-5	beaded rim jar, ridge	-	-	-	-	-	-	-	-	-	-	-	-	-	-	1	-
AM-RB-1	beaded rim bowl	-	-	-	-	-	-	1	-	-	-	1	-	-	1	-	-
AM-RB-2	carinated beaded rim bowl	1	1	-	-	-	-	1	-	-	-	2	1	1	-	-	-
AM-RC-3	incurved rim bowl	1	-	-	-	-	-	-	-	-	-	1	-	-	-	-	-

Key: Grv: grooved; Inc: incised

Appendix B

Archaeological Sites on the Deh Lurān Plain

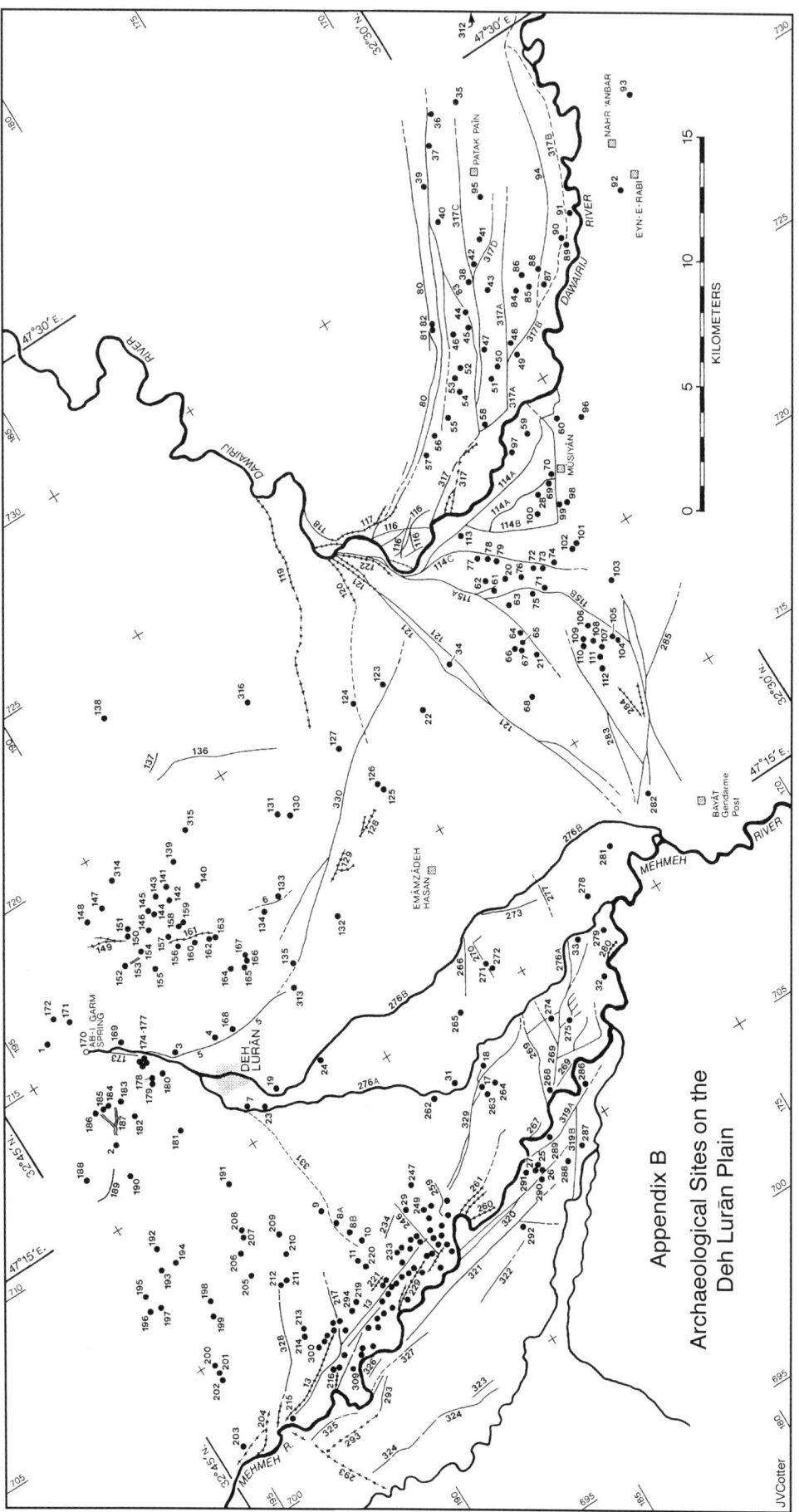

B1. Map of archaeological sites on the Deh Lurān Plain.

References Cited

Adams, Robert McC.
1962 Agriculture and urban life in early southwestern Iran. *Science* 136:109–22.
1965 *Land Behind Baghdad*. Chicago: University of Chicago Press.

Adams, Robert McC., and Hans J. Nissen
1972 *The Uruk Countryside*. Chicago: University of Chicago Press.

Aitken, Martin J.,
1998 *An Introduction to Optical Dating*. Oxford, New York: Oxford University Press.

Alizadeh, Abbas
1985 A tomb of the Neo-Elamite period at Arjan, near Behbehan. *Archëologische Mitteilungen aus Iran* 18:49–73.

Ammerman, Albert J.
1981 Surveys and archaeological research. *Annual Review of Anthropology* 10:63–88.

Boucharlat, Remy
1987 Les niveax post-achéménides à Suse, secteur nord. Fouuilles de l'Apadana-Est et de la Ville Royale-Ouest (1973–1978). *Cahiers de la Délégation Archéologique Française en Iran* 15:145–312.

Briant, Pierre
2002 *From Cyrus to Alexander: A History of the Persian Empire*. Winona Lake, Ind.: Eisenbrauns.

Brinkman, John A.
1968 *A Political History of Post-Kassite Babylonia, 1158–722 a.c.* Roma: Pontificium Institutum Biblicum.

Caldwell, Joseph R.
1968 Ghazir, Tell-i. *Reallexikon der Assyriologie* 3:348–55.

Cameron, George G.
1936 *History of Early Iran*. Chicago: University of Chicago Press.

Caran, S. Christopher, Barbara M. Winsborough, James A. Neely, and Samuel Valastro, Jr.
1995 Radiocarbon age of carbonate sediments (travertine, pedoconcretions, and biogenic carbonates): A new method based on organic residues, employing stable-isotope control of carbon sources. *Current Research in the Pleistocene* 12:75–77. Corvallis, Oregon.

Carter, Elizabeth
1971 *Elam in the Second Millennium B.C.: The Archaeological Evidence*. PhD dissertation in Near Eastern Studies, University of Chicago.
1978 Suse 'Ville Royale.' *Paléorient* 4:197–211.
1980 Excavations at Ville Royale I: The third millennium B.C. occupation. *Cahiers de la Délégation Archéologique Française en Iran* 11:11–134.
1981 Elamite ceramics. In *An Early Town on the Deh Luran Plain: Excavations at Tepe Farukhabad*, pp. 200–216. Memoirs, no. 13. Ann Arbor: Museum of Anthropology, University of Michigan.
1994 Bridging the gap between Elamites and Persians in southeastern Khuzistan. In *Achaemenid History*. Vol. VIII, edited by Heléne Sancisi-Weerdenburg, Amalie Kuhrt, and Margaret C. Root, pp. 65–95. Leiden: Nederlands Instituut voor het Nabije Oosten.

Carter, Elizabeth, and Matthew Stolper
1984 *Elam: Surveys of Political History and Archaeology*. Berkeley, CA: University of California Press.

Cook, Sherburn F., and Robert F. Heizer
1968 Relationships among houses, settlement areas, and populations in aboriginal California. In *Settlement Archaeology*, edited by K.C. Chang, pp. 79–116. Palo Alto, CA: National Press Books.

Cressey, George B.
1958 Qanats, Kariz, and Foggara. *Geographical Review* 48:17–44.
1960 *Crossroads: Land and Life in Southwest Asia*. Chicago: J.B. Lippincott.

D'Agostino, Franco
1997 *Testi Administrativi della III Dinastia di Ur dal Museum Statale Ermitage, San Pietroburgo-Russia*. Roma: Bonsignore.

Dahl, Jacob
2007 *The Ruling Family of Umma*. Istanbul: Netherlands Historisch-Archeologisch Instituut te Istanbul.

De Graef, Katrien
2005 *Les archives d'Igibuni: Les documents Ur III du chantier B à Suse*. Mémoires de la Delégation en Perse 54. Gand: L'Université de Gand.
2006 *De La Dynastie Simaski au Sukkalmahhat: Les documents fin PE IIB–début PE III du Chantier B à Suse*. Mémoires de la Delégation en Perse 55. Gand: L'Université de Gand.

Delougaz, Pinhas
1952 *Pottery from the Diyala Region*. Oriental Institute Publication 63. Chicago: University of Chicago Press.

Dewar, Robert E.
1991 Incorporating variation in occupation span into settlement-pattern analysis. *American Antiquity* 56(4):604–20.

Diehl, Michael W.
2009 Early agricultural period food provisioning and foraging. *Archaeology Southwest* 23(1):12–13.Tucson: Center for Desert Archaeology.

Dittman, René
1986 Susa in the proto-Elamite period and annotations on the painted pottery of proto-Elamite style. In *Gamdat Nasr: Period or Regional Style?* Beihefte zum Tübinger Atlas des Vorderen Orients, Series B, No. 62, edited by U. Finkbeiner and W. Röllig, pp. 332–66. Wiesbaden: Ludwig Riechert.

Dollfus, Geneviève
1978 Djaffarabad, Djowi, Bendebal: Contribution à l'étude de la Susiane au Ve millénaire et au début du IVe millénaire. *Paléorient* 4:141–67.

Doolittle, William E.
1988 *Prehispanic Occupance in the Valley of Sonora, Mexico: Archaeological Confirmation of Early Spanish Reports*. Anthropological Papers of the University of Arizona no. 48. Tucson: University of Arizona Press.

Emberling, Geoffrey
1997 Ethnicity in complex societies: Archaeological perspectives. *Journal of Archaeological Research* 5/4:295–343.

English, Paul W.
1966 *City and Village in Iran: Settlement and Economy in the Kirman Basin*. Madison: University of Wisconsin Press.
1968 The origin and spread of qanats in the Old World. *Proceedings of the American Philosophical Society* 112(3):170–81.

Fernea, Robert A.
1959 *Irrigation and Social Organization among the El Shabana: A Group of Tribal Cultivators in Southern Iraq*. PhD dissertation in Anthropology, University of Chicago.
1970 *Shaykh and Effendi: Changing Patterns of Authority among the El Shabana of Southern Iraq*. Cambridge: Harvard University Press.

Foxvog, D.A.
1980 Funerary furnishings in an Early Sumerian text from Adab. Papers read at the XXVIe Rencontre Assyriologique Internationale, edited by Bendt Alster. *Mesopotamia: Copenhagen Studies in Assyriology* 8:67–75.

Frayne, Douglas R.
1991 *The Royal Inscriptions of Mesopotamia 1: The Presargonic Period 2700–2350 B.C*. Toronto: University of Toronto Press.
1992 *The Early Dynastic List of Geographical Names*. American Oriental Series 74. New Haven, Conn.: American Oriental Society.
1993 *The Royal Inscriptions of Mesopotamia 2: Sargonid and Gutian Periods 2334–2113 B.C*. Toronto: University of Toronto Press.
1997 *Royal Inscriptions of Mesopotamia 3/2: The Ur III Period (2112–2004 B.C.)*. Toronto: University of Toronto Press.

Gasche, Hermann
1973 *La Poterie Élamite de Deuxième Millénaire a.C*. Mémoires de la Délégation Archéologique en Iran XLVII. Leiden: E.J. Brill; Paris: P. Geuthner.

Gautier, Jean-Ettienne, and G. Lampre
1905 *Fouilles de Moussian*, edited by Jacques de Morgan, pp. 59–148. Mémoires de la Délégation en Perse VIII. Paris: Ernest Leroux.

Gelb, Ignace J., and B. Kienast
1990 *Die altakkadischen Königschriften des dritten Jahrtausends v. Chr*. Freiburger altorientalische Studien 7. Stuttgart: Franz Steiner.

Gelles, Paul H.
1996 The political ecology of irrigation in an Andean peasant community. In *Canals and Communities: Small-Scale Irrigation Systems*, edited by Jonathan B. Mabry, pp. 88–115. Tucson: University of Arizona Press.

Genoulliac, Henri de
1911 *Textes cuneiform de Louvre II*. Paris: Paul Geuthner.

Gerardi, Pamela
1987 *Ashurbanipal's Elamite Campaigns*. Ann Arbor: University Microfilms.

Goetze, Albrecht
1961 Early kings of Kish. *Journal of Cuneiform Studies* 15:105–11.

Grayson, Albert K.
1963 The Walters Art Gallery Sennacherib Inscription. *Archiv für Orientforschung* 20:91:1.34.
1975 *Assyrian and Babylonian Chronicles*. Texts from Cuneiform Sources 5. Locust Valley, NY: J.J. Augustin.

Gregg, Michael W., Rhea Brettell, and Benjamin Stern
2007 Bitumen in Neolithic Iran: Biomolcular and isotopic evidence. In *Archaeological Chemistry: Analytical Techniques and Archaeological Interpretation*, edited by Michael Glasscock, Robert J. Speakman, and Rachel S. Popelka-Filcoff, pp. 137–51. American Chemical Society Symposium Series 968. New York: American Chemical Society.

Gremliza, Frederik G.L.
1962 *Ecology and Endemic Diseases in the Dez Irrigation Project*. Report to the Khuzistan Water and Power Authority and the Plan Organization of Iran. New York: Development and Resources Corporation.

Griffiths, Huw W., Antje Schwalb, and Lora R. Stevens
2001 Environmental change in southwestern Iran: The Holocene ostracod fauna of Lake Mirabad. *The Holocene* 11:757–64.

Hansman, John
1972 Elamites, Achaemenians and Anshan. *Iran* 10:101–24.

Heessel, Nils P.
2001 Pazuzu: archäologische und philologische Studien zu einem alt-orientalischen Dämon. Leiden: Brill.

Herodotus
1942 *The Persian Wars*, translated by George Rawlinson. New York: Random House.

Herr, Sarah A.
2009a The latest research on the earliest farmers. *Archaeology Southwest* 23(1):1–3. Tucson: Center for Desert Archaeology.
2009b Las Capas. *Archaeology Southwest* 23(1):9–11. Tucson: Center for Desert Archaeology.

Herrero, Pablo
1976 Tablettes administratives de Haft-Tepe. *Cahiers de la Delegation Archéologique en Iran* 6:93–116.

Herrero, Pablo, and J.J. Glassner
1990 Haft-Tepe: Choix de textes I. *Iranica Antiqua* 25:1–45.
1991 Haft-Tepe: Choix de textes II. *Iranica Antiqua* 26:39–80.

Herzfeld, Ernst
1968 *The Persian Empire. Studies in Geography and Ethnography of the Ancient Near East*. Wiesbaden: F. Steiner.

Hole, Frank
1962 Archaeological survey and excavation in Iran, 1961. *Science* 137:524–26.
1966 Investigating the origins of Mesopotamian civilization. *Science* 153:605–11.
1977 *Studies in the Archaeological History of the Deh Luran Plain: The Excavation of Chagha Sefid*. Memoirs, no. 9. Ann Arbor: Museum of Anthropology, University of Michigan.
1978 Pastoral nomadism in western Iran. In *Explorations in Ethnoarchaeology*, edited by R.A. Gould, pp. 127–67. Albuquerque: University of New Mexico Press.

Hole, Frank (editor)
1969 *Preliminary Reports of the Rice University Project in Iran 1968–1969*. Houston: Rice University, Department of Anthropology. Mimeograph.
1987 *The Archaeology of Western Iran. Settlement and Society from Prehistory to the Islamic Conquest*. Washington: Smithsonian Institution Press.

Hole, Frank, and Kent V. Flannery
1962 Excavations at Ali Kosh, Iran, 1961. *Iranica Antiqua* 2:97–148.
1968 The prehistory of southwestern Iran: A preliminary report. *Proceedings of the Prehistoric Society* 33:147–206.

Hole, Frank, Kent V. Flannery, and James A. Neely
1969 *Prehistory and Human Ecology of the Deh Luran Plain*. Memoirs, no. 1. Ann Arbor: Museum of Anthropology, University of Michigan.

Hubbs, Stephen A. (editor)
2004 *Riverbank Filtration Hydrology*. NATO Science Series. Series IV: Earth and Environmental Sciences. Dordrecht, Netherlands: Springer.

Jacobsen, Thorkild
1939 *The Sumerian King List*. Oriental Institute of the University of Chicago Assyriological Studies, no. 11. Chicago: University of Chicago Press.

Jones, Tom B., and John W. Snyder
1961 *Sumerian Economic Texts from the Third Ur Dynasty*. Minneapolis: University of Minnesota Press.

King, Leonard W.
1912 *Babylonian Boundary Stones and Memorial Tablets in the British Museum*. London: The British Museum.

Kirkby, Michael
1977 Land and water resources of the Deh Luran and Khuzistan Plains. In *Studies in the Archaeological History of the Deh Luran Plain: The Excavation of Chagha Sefid*, by F. Hole, pp. 251–88. Memoirs, no. 9. Ann Arbor: Museum of Anthropology, University of Michigan.

Kirkby, Michael J., and Anne V.T. Kirkby
1969 Provisional report on geomorphology and land use in Deh Luran and Upper Khuzistan. In *Preliminary Reports of the Rice University Project in Iran 1968–1969*, edited by Frank Hole, pp. 1–8. Houston: Rice University.

Kowalewski, Stephen A., and Suzanne K. Fish (editors)
1990 *The Archaeology of Regions*. Washington: Smithsonian Institution Press.

Kramer, Carol
1980 Estimating prehistoric populations: An ethnoarchaeological approach. In *L'Archéologie de l'Iraq du début de l'époque Néolithique à 333 avant notre ère*, edited by M.-T. Barrelet, pp. 315–34. Colloques Internationaux du Centre National de

la Recherche Scientifique no. 580. Paris: Editions du Centre National de la Recherche Scientifique.

Larsen, Curtis E.
1983 *Life and Land Use on the Bahrain Islands.* Chicago: University of Chicago Press.

LeBlanc, Stephen A.
1971 An addition to Naroll's suggested floor area and settlement population relationship. *American Antiquity* 36:210–11.

Legrain, Léon
1912 *Le temps des rois d'Ur: recherches sur la société antique d'après des textes nouveaux.* Paris: Champion.

Lorimer, D.L.R.
1908 *A Report on the Pusht-i Kuh.* Account of Luristan, in the possession of the author.

Luckenbill, Daniel D.
1924 *The Annals of Sennacherib.* Oriental Institute Publications 2. Chicago: University of Chicago Press.

Mabry, Jonathan B. (editor)
2008 *Las Capas: Early Irrigation and Sedentism in a Southwestern Floodplain.* Center for Desert Archaeology, Anthropological Papers 28. Tucson: University of Arizona Press.

Marschner, Robert F., Leo J. Duffy, and Henry T. Wright
1978 Asphalts from ancient townsites in southwestern Iran. *Paleorient* 4:97–112.

Masse, W. Bruce
1981 Prehistoric irrigation systems in the Salt River Valley, Arizona. *Science* 214:408–15.

McMahon, Augusta
2006 *Nippur V. The Early Dynastic to Akkadian Transition: The Area WF Sounding.* Oriental Institute Publications 129. Chicago: Oriental Institute of the University of Chicago.

Michalowski, Piotr
1983 History as charter: Some observations on the Sumerian King List. *Journal of the American Oriental Society* 103/1:237–48.
2003 A man called Enmebaragesi. In *Literatur, Politik, und Recht in Mesopotamien,* edited by Walther Sallaberger, Konrad Volk, and Annette Zgoll, pp. 195–208. Orientalia Biblica et Christiana. Heidelberg: Harrassowitz.
2008 Observations on "Elam" and "Elamites" in Ur III times. In *On the IIIrd Dynasty of Ur: Studies in Honor of Marcel Sigrist,* edited by Piotr Michalowski. Boston: American Schools of Oriental Research.
in press Šimaški. *Reallexicon der Assyriologie.*

Miroschedji, Pierre de
1981a Fuilles du chantier Ville Royale II à Suse (1975–1977). I. Les niveaux élamite Archeologique. *Cahiers de la Délégation Archéologique Française en Iran* 12:9–136.
1981b Prospections archéologiques au Khuzistan en 1977. *Cahiers de la Délégation Archéologique Française en Iran* 12:169–92.
1986 La localisation de Madaktu et l'organisation de l'Elam à l'époque Neo-Elamaite. In *Fragmenta Historica Elamitica,* edited by Leon de Meyer, Hermann Gasche, and François Vallat, pp. 209–25. Paris: Editions Recherche sur les Civilisations.
1987 Fouille du chantier Ville Royale II à Suse (1975–1977). II. Niveaux d'époques achéménide parthe et islamique. *Cahiers de la Délégation Archéologique Française en Iran* 15:11–144.
1990 La fin de l'Elam: Essai d'analyse et d'interprétation. *Iranica Antiqua* 25:47–95.

Muscarella, Oscar White
1985 Review of: Nush-i Jan III: The small finds, by John Curtis. *Journal of the American Oriental Society* 105:729–30.

Naroll, Raoul
1962 Floor area and settlement population. *American Antiquity* 27:587–89.

Neely, James A.
1974a *The Prehistoric Lunt and Stove Canyon Sites, Point of Pines, Arizona.* PhD dissertation in Anthropology, University of Arizona, Tucson. Ann Arbor: University Microfilms.
1974b Sassanian and early Islamic water-control and irrigation systems on the Deh Luran Plain, Iran. In *Irrigation's Impact on Society,* edited by Theodore E. Downing and McGuire Gibson, pp. 21–42. Anthropological Papers of the University of Arizona no. 25. Tucson: University of Arizona Press.
1995 Mogollon/Western Pueblo soil and water control systems of the Reserve phase: New data from west-central New Mexico. In *Soil, Water, Biology, and Belief in Prehistoric and Traditional Southwestern Agriculture,* edited by H.W. Toll, pp. 239–62. Special Publication 2. Albuquerque: New Mexico Archaeological Council.
2005 Prehistoric agricultural and settlement systems in Lefthand Canyon, Safford Valley, southeastern Arizona. In *Inscriptions: Papers in Honor of Richard and Nathalie Woodbury,* edited by R.N. Wiseman, T. O'Laughlin, and C.T. Snow, pp. 145–69. Papers of the Archaeological Society of New Mexico, No. 31.
2010a Prehistoric water management in highland Mesoamerica. In *Water and Humanity: Historic Overview,* edited by V.L. Scarborough and Y. Yasuda. *The UNESCO History of Water and Civilization,* Vol. VII. New York: Springer, in press.
2010b Parthian and Sasanian settlement patterns of the Deh Luran Plain, Khuzistan Province, southwestern Iran. In *Advances in Parthian and Sasanian Studies,* edited by St. John Simpson. Acta Iranica. Leuven, Belgium: Peeters, Ltd., in press.
2010c Prehistoric agricultural strategies in the Safford Basin, southeastern Arizona. In *Between Mimbres and Hohokam: Exploring the Archaeology and History of Southeastern Arizona and Southwestern New Mexico,* edited by H.D. Wallace. Amerind Foundation Archaeological Series, No. 12. Tucson: University of Arizona Press, in press.

Neely, James A., S. Christopher Caran, and Barbara M. Winsborough
1990 Irrigated agriculture at Hierve el Agua, Oaxaca, Mexico. In *Debating Oaxaca Archaeology,* edited by J. Marcus, pp. 115–89. Anthropological Papers, no. 84. Ann Arbor: Museum of Anthropology, University of Michigan.

Neely, James A., and Everett J. Murphy
2008 Prehistoric Gila River canals of the Safford Basin, southeastern Arizona: An initial consideration. In *Crossroads of the Southwest: Culture, Ethnicity, and Migration in Arizona's Safford Basin.* Proceedings of the Arizona Archaeological Council (AAC) Fall 2005 meeting, edited by David E. Purcell. Newcastle upon Tyne, United Kingdom: Cambridge Scholars Publishing.

Neely, James A., and Carlos A. Rincón Mautner
2004   *Los Canales "Fosilizados" del Valle de Tehuacan: Un Proyecto Arqueológico con Levantamiento Cartográfico y Recolección Multidisciplinaria de Muestras.* Informe Final al Consejo de Arqueología del Instituto Nacional de Antropologia e Historia de México. México, D.F.

Neely, James A., and Henry T. Wright
1994   *Early Settlement and Irrigation on the Deh Luran Plain: Village and Early State Societies in Southwestern Iran.* Technical Report, no. 26. Ann Arbor: Museum of Anthropology, University of Michigan.

Neely, James A., Henry T. Wright, and others
forthcoming   *The Parthian, Sasanian, and Early Islamic Settlements of the Deh Luran Plain, Southwestern Iran.* Ann Arbor: Museum of Anthropology, University of Michigan.

O'Brien, Michael J., Roger D. Mason, Dennis E. Lewarch, and James A. Neely
1982   *A Late Formative Irrigation Settlement Below Monte Alban: Survey and Excavation on the Xoxocotlan Piedmont, Oaxaca, Mexico.* Institute of Latin American Studies. Austin: University of Texas.

Piepkorn, Arthur C.
1933   *Historical Prism Inscriptions of Ašurbanipal I: Editions E, B 1-5, D and K.* Assyriological Studies 5. Chicago: University of Chicago Press.

Plog, Fred T.
1973   Diachronic anthropology. In *Research and Theory in Current Archaeology*, edited by C.L. Redman, pp. 181–98. New York: John Wiley.

Plog, Steven, and J.L. Hantman
1990   Chronology construction and the study of prehistoric culture change. *Journal of Field Archaeology* 17:439–56.

Postgate, John Nicholas
1992   *Early Mesopotamia*. London: Routledge.

Postgate, John Nicholas, and R.A. Mattila
2004   Il-Yada' and Sargon's southeast frontier. In *From the Upper Sea to the Lower Sea: Studies on the History of Assyria and Babylonia in Honour of A.K. Grayson*, pp. 235–54. Istanbul: Nederlands Instituut Voor et Nabije Oosten.

Potts, Daniel T.
1999   *The Archaelogy of Elam*. Cambridge: Cambridge University Press.

Rawlinson, Major Henry
1839   Notes on a march from Zohab, along the mountains to Khuzistan (Susiana) and from thence through the province of Luristan to Kirmanshah in the year 1836. *Journal of the Royal Geographical Society of London* IX:26–116.

Ray, Chittaranjan, Gina Melin, and Ronald B. Linsky (editors)
2002   *Riverbank Filtration: Improving Source-Water Quality*. Water Science and Technology Library. Dordrecht, Netherlands: Kluwer Academic.

Reade, Julian
1983   *Assyrian Sculpture*. London: The British Museum.

Redding, Richard W.
1978   Rodents and the archaeological paleoenvironment: Considerations, problems and the future. In *Approaches to Faunal Analysis in the Middle East*, edited by R.H. Meadow and M.A. Zeder, pp. 63–68. Cambridge, MA: Peabody Museum of Archaeology and Ethnology.
1981   The faunal material. In *An Early Town on the Deh Luran Plain: Excavations at Tepe Farukhabad*, pp. 233–61, 391–425. Memoirs, no. 13. Ann Arbor: Museum of Anthropology, University of Michigan.

Reiner, Erica
1973   The location of Anšan. *Revue d'Assyriologie* 67:57–62.

Rowton, Michael B.
1960   The date of the Sumerian King List. *Journal of Near Eastern Studies* 19/2:156–62.

Sajjadi, Mansour Seyyed
1982   *Qanat/Kariz: Storia, tecnica construttiva ed evoluzione*. Teheran: Instituto Italiano di Cultura.

Sauren, Herbert
1969   *Wirtschafturkunden aus der Zeit der III. Dynastie von Ur im Besitz des Musée d'Art et d'Histoire in Genf*. Naples: Istituto Orientale di Napoli.

Schacht, Robert M.
1984   The contemporaneity problem. *American Antiquity* 49(4):678–95.

Scheil, Vincent
1907   Textes élamite-anzanite, troisième série. *Mémoires de la Délégation en Perse, Tome IX*. Paris: Le Roux.
1913   Textes élamites-sémitiques. *Mémoires de la Délégation en Perse, Tome XIV*. Paris: Le Roux.
1939   Mélanges épigraphiques. *Mémoires de la Mission Archéologiques en Perse, Tome XXVIII*. Paris: Le Roux.

Selz, Gebhard J.
1989   *Altsumerische Verwaltungstext aus Lagaš*. Frieburger Altorinetalische Studien, Band 15/1. Stuttgart: Steiner.
1991   "Elam" und "Sumer." In *Mesopotamie et Elam*, edited by Leon de Meyer and Hermann Gasche, pp. 27–43. Mesopotamian History and Environment, Occasional Publications No. 1. Ghent: Recherches et Publications.

Sigrist, Marcel
1988   *Neo-Sumerian Account Texts in the Horn Archaeological Museum*. Berrien Springs, Michigan: Andrews University Press.
1990   *Messenger Texts from the British Museum*. Potomac, Maryland: Capital Decisions.
1993   *Texts from the British Museum*. Sumerian Archival Texts I. Bethesda, Maryland: CDL Press.

Sollberger, Edmond
1956   *Corpus des inscriptions "royales" présargoniques de Lagaš*. Genève: E. Droz.

Southworth, C.H.
1931 *Gila River Survey Report*. Vol. 1, *Supplemental Exhibits*. Manuscript on file at the Pueblo Grande Museum, Phoenix, Arizona.

Stanbury, Pamela C.
1996 The utility of tradition in Sri Lankan bureaucratic irrigation: The case of the Kirindi Oya project. In *Canals and Communities: Small-Scale Irrigation Systems*, edited by Jonathan B. Mabry, pp. 210–26. Tucson: University of Arizona Press.

Steinkeller, Piotr
1980 On the reading and location of the toponyms ÚR×Ú.KI and A.HA.KI. *Journal of Cuneiform Studies* 32/1:23–33.
1982 The question of Marḫaši: A contribution to the historical geography of Iran in the third millennium B.C. *Zeitschrift für Assyriologie* 72:237–65.
2003 An Ur III manuscript of the Sumerian King List. In *Literatur, Politik und Recht in Mesopotamien: Festschrift für Claus Wilcke*, edited by Walter Sallaberger, Konrad Volk, and Annette Zgoll. Wiessbaden: Harrassowitz.
2007 New light on Šimaški and its rulers. *Zeitschrift für Assyrologie und vorderasiatische Archäologie* 97/2:215–32.

Stève, Marie-Joseph, and Hermann Gasche
1971 *L'Acropole de Suse*. Mémoires de la Délégation Archéologique en Iran 46. Paris: Geuthner.

Stevens, Lori R., Emi Ito, Antje Schwalb, and Herbert E. Wright, Jr.
2006 Timing of atmospheric precipitation in the Zagros Mountains inferred from a multi-proxy record from Lake Mirabad, Iran. *Quaternary Research* 66:494–500.

Stolper, Matthew W.
1982 On the dynasty of Šimaški and the early Sukkalmaḫḫu. *Zeitschrift für Assyriologie* 72:42–67.
1984 Political history. In *Elam: Surveys of Political History and Archaeology*, by Elizabeth Carter and Matthew W. Stopler, pp. 3–100. University of California Publications, Near Eastern Studies No 25. Berkeley: University of California Press.

Streck, Maximillian
1916 *Assurbanipal und die letzten assyrischen könige bis zum untergange Niniveh's*. Leipzig: J.C. Hinrichs.

Sumner, William M.
1974 Excavations at Tal-I Malyan 1971–1972. *Iran* 12:155–80.

Szarzyńska, Krystyna
1996 Archaic Sumerian standards. *Journal of Cuneiform Studies* 48:1–15.

Thureau-Dangin, François
1910 *Inventaire des Tablettes de Telloh*. Vol. 1. Paris: Le Roux.

Tolstoy, Paul
1975 Settlement and population trends in the Basin of Mexico (Ixtapaluca and Zacatenco phases). *Journal of Field Archaeology* 2:97–104.

Vallat, François
1980 *Suse et l'Elam*. Recherches sur les Grands Civilsations. Association pour la diffusion de la pensée Française. Paris: Editions du CNRS.
1993 *Les noms géographiques des sources suso-élamites*. Repertoire Géographiques des Textes Cunéiformes 11. Weisbaden: F. Reichert.

Vanden Berghe, Louis
1968 La nécropole de Bani Surmah. *Archéologia* 24:53–63.
1970 La nécropole de Kalleh Nissar. *Archéologia* 32:64–73.

Van Dijk, J.
1978 Ishbierra, Kindattu, l'homme d'Elam, et la chute d'Ur. *Journal of Cuneiform Studies* 30:189–208.

Van Zeist, Willem, and Sitse Bottema
1977 Palynological investigations in western Iran. *Palaeohistoria* XIX:19–85.

Waters, Matthew W.
2000 *A Survey of Neo-Elamite History*. State Archives of Assyria Studies XII. Helsinki: Institute for African and Asian Studies, University of Helsinki.

Watson, Richard A., and Herbert E. Wright, Jr.
1969 The Saidmarreh landslide, Iran. *Geological Society of America, Special Paper* 123:115–39.

Weiss, Harvey
1977 Periodization, population and early state formation in Khuzistan. In *Mountains and Lowlands: Essays in the Archaeology of Greater Mesopotamia*, edited by L.D. Levine and T.C. Young, Jr., pp. 347–69. Malibu, CA: Undena.

Wilke, C.
1987 Inschriften 1983–1984. In *Isin-Ishan Baḥriyāt III*, edited by B. Hrouda, pp. 83–120. Philosophische-historische Klasse. Abh N.F. 94. Munich: Bayerische Akademie der Wissenschaften.

Wilkinson, T.J.
2003 *Archaeological Landscapes of the Near East*. Tucson: University of Arizona Press.

Winsborough, Barbara M., S. Christopher Caran, James A. Neely, and Samuel Valastro, Jr.
1996 Calcified microbial mats date prehistoric canals: Radiocarbon assay of organic extracts from travertine. *Geoarchaeology* 11(1):37–50. New York: John Wiley and Sons Publishers.

Woodbury, Richard B., and James A. Neely
1972 Water control systems of the Tehuacan Valley. In *The Prehistory of the Tehuacan Valley*. Vol. 4, *Chronology and Irrigation*, edited by R.S. MacNeish et al., pp. 81–153. Austin: University of Texas Press.

Wright, Henry T.
1981 The southern margins of Sumer. In *Heartland of Cities*, by Robert McC. Adams, pp. 295–345. Chicago: University of Chicago Press.

Wright, Henry T. (editor)
1981 *An Early Town on the Deh Luran Plain: Excavations at Tepe Farukhabad*. Memoirs, no. 13. Ann Arbor: Museum of Anthropology, University of Michigan.

Wright, Henry T., and Gregory A. Johnson
1975 Population, exchange and early state formation in southwestern Iran. *American Anthropologist* 77:267–89.

Young, T. Cuyler, Jr.
1967 The Iranian migration into the Zagros. *Iran* 5:11–34.

1969 *Excavations at Godin Tepe: First Progress Report*. Royal Ontario Museum, Art and Archaeology, Occasional Paper 17. Toronto: Royal Ontario Museum.

Zadok, R.
1994 Elamites and other peoples from Iran and the Persian Gulf region in early Mesopotamian sources. *Iran* 32:31–51.

The Deh Lurān Plain looking northeast from Farukhābād (DL-32) in 1961. Note high proportion of goats in the nearby herd (photograph by F. Hole).

New road on the Deh Lurān Plain in spring, 1968 (photograph by H. Wright).

The Deh Lurān Plain looking northwest from Musiyān (DL-20). DL-61 in the center foreground is a small site of the IVth millennium B.C. (photograph by J. Neely).

Jim Neely recording rock features in 1969 (photograph by F. Hole).

Plate 1. Deh Lurān landscapes.

Musiyān in 1969 from the west (photograph by J. Neely).

Musiyān (DL-20) in 1961 (courtesy World Wide Surveys).

Musiyān about 2006. The rectangular ditch of the French field camp oriented to the quarter points is south of the center of the mound. 100 m north are the public buildings. Traces of the town's walls are evident on the west and north sides. The scale is 100 m (courtesy Google Earth/Digital Globe).

Plate 2. Images of Musiyān (DL-20).

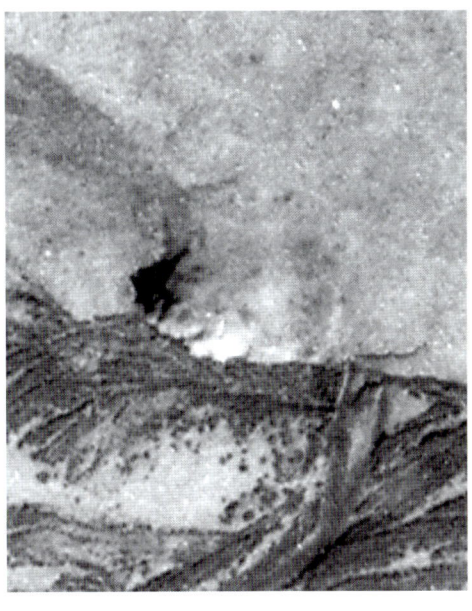

Farukhābād (DL-32) in 1961 (courtesy World Wide Surveys).

Farukhābād (DL-32) from the northwest in spring of 1968 (photograph by H. Wright).

Farukhābād about 2006. Two black goat hair nomad tents and associated corrals are visible on the eastern lower terrace of the mound. The line of Excavation C is the light line visible in the northeast side of the high mound. The northeast corner of the lower mound has been leveled to facilitate agriculture. The scale is 100 m (courtesy Google Earth/Digital Globe).

Plate 3. Images of Farukhābād (DL-32).

Gārān (DL-34) in 1961 (courtesy World Wide Surveys).

Gārān and Kuh-i Siyāh from Musiyān, 1963 (photograph by F. Hole).

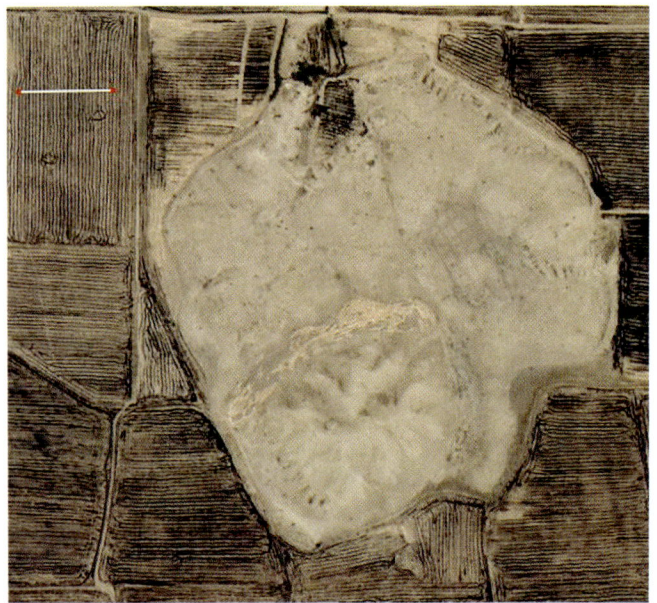
Gārān about 2006. Note the low area, a possible moat, around the high southern mound. No architecture is visible on this image. The dark marks on the northern perimeter are probable tank emplacements. The scale is 100 m (courtesy Google Earth/Digital Globe).

Plate 4. Images of Gārān (DL-34).

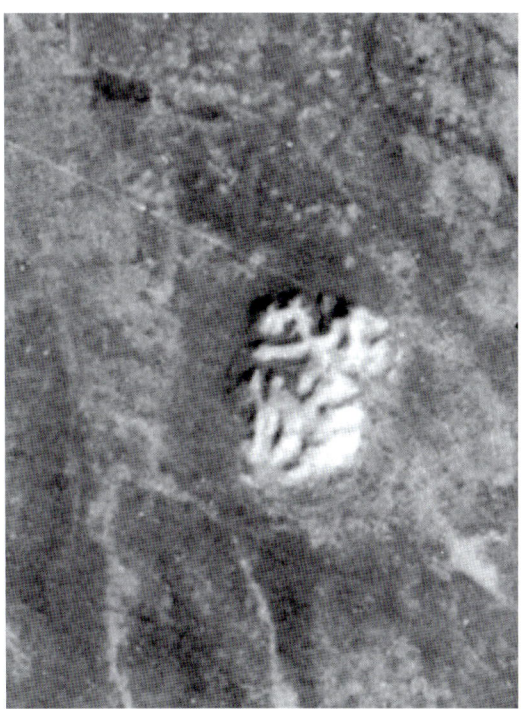

Tepe Patak (DL-35) in 1961 (courtesy World Wide Surveys).

Tepe Patak about 2006; the scale is 100 m (courtesy Google Earth/Digital Globe).

Patak (DL-35) was a small Early Dynastic site on the southeast plain reoccupied as a major settlement in Middle Elamite, Neo-Elamite, and Achaemenid times. On the 1961 image one can see small mounds, which are at present undated, southeast and north of the site.

Plate 5. Images of Patak (DL-35).

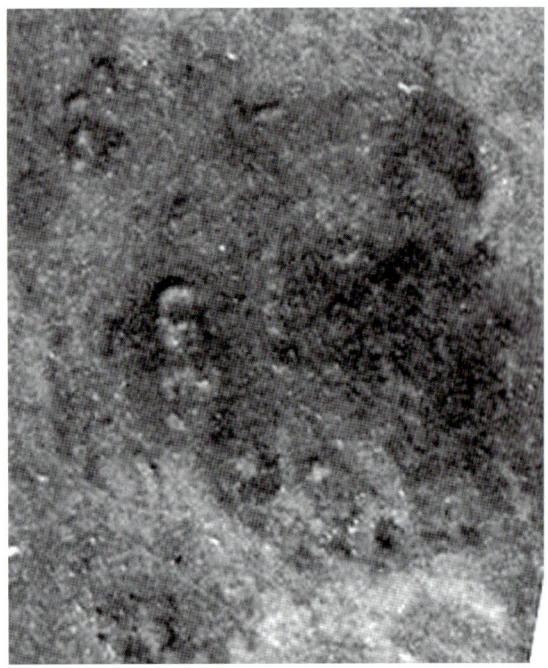 

DL-41 in 1961; note dark rectangle in middle (courtesy World Wide Surveys).

DL-41 about 2006; the scale is 100 m (courtesy Google Earth/Digital Globe).

A large unnamed Early Dynastic site, DL-41 has an oval or roughly rectangular wall, within which many small mounds are evident. The large mounds on the west edge have later material. The dark rectangular area in the middle could be public buildings.

New canal near Gārān (DL-34) in March. Note the riparian vegetation and salt on canal bank.

Plate 6. Images of DL-41, and new canal near Gārān.

Railroad on canal (DL-331) cutting Tāfuleh (DL-8) in 1969. Note the cobble footing of a rectangular structure (photograph by J. Neely).

Eroded shafts of a qanāt (DL-13) west of Deh Lurān in 1969 (photograph by J. Neely).

Plate 7. Deh Lurān water management features.

Moldmade clay bed scenes from Tepe Patak (DL-35). *upper*, Plaque A; *lower*, Plaque B (scales in cm).

DL-103
Pazuzu figurine
(scale in cm)

Plate 8. Human and divine representations from the Deh Lurān Plain.